Costing Improved Water Supply Systems for Low-income Communities

Costing Improved Water Supply Systems for Low-income Communities

A Practical Manual

Fabrizio Carlevaro and Cristian Gonzalez

IWA PUBLISHING

Published by

IWA Publishing
Alliance House
12 Caxton Street
London SW1H 0QS, UK
Telephone: +44 (0)20 7654 5500
Fax: +44 (0)20 7654 5555
Email: publications@iwap.co.uk
Web: www.iwapublishing.com

First published 2015

British Library Cataloguing in Publication Data
A CIP catalogue record for this book is available from the British Library

ISBN: 9781780407210 (Paperback)
ISBN: 9781780407227 (eBook)

Contents

List of figures

Downloads

Companion tools to the book can be downloaded from the following locations.

http://www.iwawaterwiki.org/WSCP.xls

http://www.iwawaterwiki.org/WSCP_Guantanamo-San%20Martin.xls

About the authors

Dr. Fabrizio Carlevaro is emeritus professor of the Faculty of Economic and Social Sciences of the University of Geneva (now Geneva School of Economics and Management, Switzerland). He spent most of his academic career at the Department of Econometrics of this university teaching theoretical and applied econometrics, engineering-economic modelling and survey sampling techniques. His applied research activity encompasses: system demand analysis, energy and environmental issues and policies, income tax design and forecasting, water and sanitation issues. On these subjects, he has been called as adviser by Swiss cantonal and federal authorities, and international organizations.

Dr. Cristian Gonzalez is an econometrician and economist interested in the evaluation of environmental, microfinance and public projects. He worked as economist and statistician at the Inter-American Development Bank, the International Labour Organization and for consulting firms on applied economics and data assessment. He joined the research project 'Development of a methodology to estimate the costs of improving access to safe water supply and sanitation facilities at a national level' as an assistant of the Department of Econometrics and later as a scientific collaborator of the Laboratory of Applied Economics of the University of Geneva. He is presently Director of Statistics and Data Analysis of the International Road Federation in Geneva.

Foreword

This Manual which accompanies the tool for costing improved water supply systems for low-income communities is published by IWA in the auspicious year 2015. The period of the Millennium Development Goals is drawing to an end. In 2000, world leaders signed up to a vision for development when they adopted the Millennium Declaration. We now know what this vision has brought to bear.

Certainly for drinking-water supply the global impact of the MDGs has been significant. The 2014 update report of the WHO/UNICEF Joint Monitoring Programme for Water Supply and Sanitation states that since 1990 well over two billion people gained access to improved sources of drinking-water, that by 2012 89% of the world's population used improved sources and that 56% (almost four billion people) now enjoy a piped drinking-water connection on their premises.

This Manual is very much a brainchild of the MDG period. Its authors, both econometricians at the University of Geneva, Switzerland, were inspired by the information on access to drinking-water emerging from household surveys and presented in the JMP reports. They realized from JMP's disaggregation of datasets that the global figures conceal huge discrepancies: between different regions of the world, between rural and urban populations and, in cities, between formal municipalities and the surrounding informal settlements.

When the Water, Sanitation and Health unit of the World Health Organization contacted Professor Carlevaro to explore options for collaboration in the field of WASH economics, both sides rapidly recognized the opportunities resulting from their complementary areas of expertise.

In the ensuing discussions, the authors rapidly caught up with the nature and magnitude of the challenges faced by efforts to expand drinking-water supply coverage. And they learned that for the marginalized, the under-privileged, for those population groups with the least access to improved water resources, the solutions were not simply confined to infrastructural development and technical improvements.

Just like any other allocation of public resources, investment in water supply for low-income communities needs a proper evidence base. These are communities of people who live in remote, often water-scarce areas or in unplanned peri-urban shantytowns with extreme population densities. Yet, decision-makers need to know about the costs of solution options before they can proceed with informed investment planning.

Full benefit-cost analyses are, however, daunting tasks for engineers dealing with drinking-water supply for these communities. And they cost a disproportionate amount of money.

The philosophy behind the tool and Manual is to focus on the cost part of these analyses. In the specific context of low-income communities this will provide sufficient evidence for the selection from a limited number of options. The tool developed by Fabrizio Carlevaro and Cristian Gonzalez takes the user in a systematic, step-by-step fashion through the cost analysis, relying in part on the reader's general knowledge of budget planning and adding the essential elements of economic analysis to arrive at a reliable, realistic and acceptable outcome.

When WHO organized two training workshops in South East Asia as part of testing the tool, I was pleased to note that the Manual struck a chord with the intended users. Not only did they engage enthusiastically in testing the tool and Manual in real-life situations in the participating six countries and in reporting back at the second workshop, but in the wake of these events they continued their communications with the authors, showing their interest in applying the material they received and the value of its outcome.

It also became clear from these contacts with the targeted readership that there is considerable interest in a similar tool to cost sanitation options for low-income communities. Currently, the authors are exploring options to respond to this need.

The MDG period is over, the new era of Sustainable Development Goals is about to start, with a broader set of water and sanitation targets. Yet the challenge of deciding about resource allocation for drinking-water supply projects remains. The tool and Manual can perform an important role in the outreach to those communities that have been left behind and they will provide strong support to efforts to achieve universal coverage by 2030.

Robert Bos
Senior Advisor
International Water Association
Geneva, February 2015

Preface

This manual and costing tool have been prepared in response to a need identified by the Water, Sanitation and Health Programme (WSH) of the World Health Organization (WHO). The Organization was faced with the challenge to contribute in a meaningful way to achieving one of the main targets of the Millennium Development Goals, namely to halve, by 2015, the proportion of people without sustainable access to safe drinking water and basic sanitation. It decided to tackle the costing of options for improved access, both to safe drinking water and to adequate sanitation. A questionnaire was sent to all WHO Member States in 2000, enquiring about the costs of drinking water and sanitation improvements. Despite a successful response rate, the outcome of this survey was unable to provide reliable cost estimates.

It was concluded that a more robust method of cost analysis was required. WHO-WSH approached the former Department of Econometrics of the University of Geneva with the suggestion to jointly develop a project proposal on this issue, to be submitted for funding to the Geneva International Academic Network (GIAN). At the time, GIAN was an international research network founded by the University of Geneva, the Graduate Institute of International Studies and the Graduate Institute of Development Studies, to create synergies among academic institutions and international organizations to further the role of Geneva and Switzerland in the service of peace and justice. This research project was formulated in 2003 with the title 'Development of a methodology to estimate the costs of improving access to safe water supply and sanitation facilities at a national level'. With the support from GIAN through a so-called 'Small Grant', it got under way in 2004. The project was intended to result in criteria and procedures that would ensure achievement of maximum community health benefits at minimum cost of investing in improved water supply and sanitation technologies.

Although limited in scope to the process of costing safe water supply and sanitation technologies, our project was intended to be part of a more comprehensive social valuation of basic services that are instrumental in fostering human development and quality of life in developing countries. For this reason, a proper use of our method must lie within a larger setting taking into account, comprehensively, the cultural, environmental, institutional, political and social conditions that should be used by policy decision makers in developing countries to promote sustainable development strategies.

The present manual and tool provide practical guidance to facilitate and standardize the implementation of social life-cycle costing to 'improved' drinking-water supply technologies. These technologies have been defined by the WHO/UNICEF Joint Monitoring Programme for Water Supply and Sanitation, as those that, by the nature of its construction, adequately protect the source of water from outside contamination, in particular with faecal matter. While the conceptual framework we use has also been conceived to be applied to sanitation technologies, it was decided to address the elaboration of practical guidance for costing improved sanitation options in a separate manual and tool, still in preparation.

In order to cost – from a social point of view – an improved water supply technology, an analytical approach is used that disaggregates the technology process according to its essential components, singled out by an engineering description. To support the implementation of the method, questionnaires have been designed to identify the main resources invested in a water supply project and to collect, at different disaggregation levels, investment and recurrent costs incurred during the expected life-cycle of the technology. We have also designed procedures for processing, aggregating and interpreting these data that will help identify least-cost options and contribute the cost components of a least-cost analysis, the aim being to promote the most efficient allocation of resources in terms of cost per unit of service provided. To facilitate the application of our costing method to actual projects, a basic tool was developed using Microsoft Excel, which we call a water supply costing processor (please see page x for the download links). It enables a user-friendly implementation of all the tasks involved in a social life-cycle costing process and provides both the detailed and the consolidated cost figures that are needed by decision-makers.

The rationale of this costing method and the practical use of the Excel costing tool are presented in detail in this manual. For a more concise presentation of its contents, one can refer to Carlevaro and Gonzalez (2011).

Acknowledgements

Many people deserve to be acknowledged for their intellectual contribution to the conception and realization of this project. Our main collaborator, Dr. Cristian Gonzalez, has been the enthusiastic and persevering pillar of the project, in charge of compiling the practical manual and programming the water supply costing processor under my supervision. Without him this project couldn't have been achieved. At the beginning of the project, Marie-Anna Leclerc, was charged to collect and organize relevant data on water supply and sanitation technologies during her four-month training period at the Department of Econometrics. This was carried out as part of the thesis work for her master degree in applied quantitative economics of the University of Reunion Island. Later, architect Nadia Carlevaro helped us to integrate all this technical information into our manual, shaping its present format. The WHO team involved in the project, namely Jamie Bartram, Laurence Haller, Jose Hueb and Bernard Dizier as well as Pete Kolsky of the World Bank, have been of invaluable help in preparing the initial project submitted for funding to GIAN and in commenting and criticizing our methodological choices. In particular, Laurence devoted a lot of time to establishing the useful links with the circle of water supply and sanitation professionals. These links allowed the improvement of the guidance provided by our manual by integrating into it the valuable comments and suggestions formulated by development economist Guy Hutton and by Terrence Thompson, Regional Adviser Environmental Health at the WHO Regional Office for the Western Pacific. The meticulous and insightful editing work by Angela Haden has greatly improved the clarity of this final version of the manual. Last but not least, I wish to address special thanks to Robert Bos, former Coordinator of the WHO Water, Sanitation, Hygiene and Health Unit for his enthusiastic help and everlasting support in achieving the completion of this project, introducing its results to potential users and finally in making possible the dissemination of our manual and costing tool through IWA Publishing.

Fabrizio Carlevaro
Project Leader
Emeritus Professor
Geneva School of Economics and Management
University of Geneva, Switzerland

Chapter 1

Background and objectives

1.1 WATER AND HEALTH

Poor water quality and sanitation continue to pose a major threat to human health. Diarrhoea spreads easily in an environment of poor hygiene and inadequate sanitation, by killing about 2.2 million people each year, most of them children younger than 5 years of age (WHO/UNICEF, 2000). Some 200 million people worldwide are infected with schistosomiasis, of whom 20 million suffer severe consequences, and 146 million people are threatened with blindness resulting from trachoma.

Water quality is deteriorating in many places, and some cities in the developing world treat only about 10% of their sewage. As a result, developing countries are facing a health crisis of large proportions.

1.1.1 The United Nations Millennium Development Goals

The United Nations Millennium Declaration confirmed the central role of water in sustainable development and in efforts to eradicate poverty (General Assembly resolution 55/2 of 8 September 2000, United Nations Millennium Declaration). One of the key targets stated in the Millennium Development Goals derived from that Declaration is to halve, by 2015, the proportion of people without sustainable access to safe drinking-water. In addition, the 2002 World Summit on Sustainable Development extended the target to include access to sanitation, namely to halve, by 2015, the proportion of people who do not have access to basic sanitation.

Since 1990, WHO and UNICEF have managed a Joint Monitoring Programme for Water Supply and Sanitation (WHO/UNICEF, 2010). The overall aim of the Joint Monitoring Programme (JMP) is to report globally on drinking-water supply and sanitation. In particular, the JMP has been formally designated to monitor progress in achieving the water supply and sanitation targets of the Millennium Development Goals (MDG).

The estimates presented in the 2012 JMP report (WHO/UNICEF, 2012), describing the situation as of end-2010, brought the news that the MDG for drinking-water had been already met in 2010, five years ahead of schedule; but warned that despite this enormous accomplishment the job was far from finished, as 780 million people (mainly located in rural areas) lacked of improved drinking-water sources. Although this trend was confirmed by the 2013 JMP report (WHO/UNICEF, 2013), in a recent article Cumming *et al.* (2014) have challenged the method used by JMP for monitoring the progress in accessing to safe drinking-water. According to these authors, although there is a single MDG target for water and sanitation, the respective benchmarks, set by the JMP for water and sanitation, differ significantly. Indeed, the benchmark for access to improved drinking-water sources is defined at community-level, while for sanitation it is at household-level. Re-estimating the global progress in accessing to improved drinking-water sources at household-level shows that 2015 MDG target for safe drinking-water access has not been achieved in 2010, and will not been reached at the 2015 target date. These findings stress the urgent need to invest in improved sources of drinking-water for rural and suburban communities, which have the largest number of individuals deprived of this basic resource.

1.1.2 Health benefits of safe water and basic sanitation

Studies (WHO, 2004) show that better access to safe water and basic sanitation along with the practice of hygiene, have significant impacts on health.

- Safe water supply can reduce episodes of diarrhoea by 21%.
- Basic sanitation can reduce episodes of diarrhoea by 37.5%.
- Providing a piped water supply to the household, with treatment to remove pathogens and quality monitoring, as well as a sewage connection with partial treatment of wastewaters (as exists in developed countries) could achieve a reduction in diarrhoeal episodes in developing countries of up to 77%.

There are also important development and livelihood opportunities that are affected by investment in water supply, sanitation and hygiene. These include reduced costs for the health sector, reduced costs of health care at the household level, reduced opportunity costs (particularly arising from less time spent on collecting water), and improved school attendance.

Expanding access to safe water and basic sanitation for those who currently lack such access would do much to reduce the global burden of water-related diseases and to improve the quality of life. The question is: how much would this cost?

1.1.3 Previous WHO work

Costs consist of all resources required to put in place and maintain an intervention, including investment costs (such as planning, construction, and house alterations) and recurrent costs (operation, maintenance, monitoring, regulation, and so on).

Hutton and Haller (2004) estimated the total global costs per year to achieve a selected number of targets. The main source of data was a survey report by WHO/UNICEF (2000). All WHO Member States were sent a questionnaire asking about the costs of construction for water and sanitation improvements, at cost per person covered.

The response rate was high. The authors felt, however, that some of the data were unreliable, but time and resources were insufficient to follow up or validate the data. Assumptions were made about the working life of the equipment involved in making the improvements, and about the costs of operation and maintenance, monitoring, regulation, and other related activities, as a percentage of the annual equivalent investment cost. A final total cost per intervention, per person covered, per year was estimated.

The method used was satisfactory for a global analysis. For application at national and local levels, however, a method is needed that offers greater accuracy, sensitivity and specificity. At the same time, a method to be applied on a smaller scale should be more affordable, to make the possibility of actually using it more realistic.

Joint research by WHO and the Department of Econometrics of the University of Geneva focused on the process of cost estimation. The product of this research, a new method that is compatible with the global economic evaluation, is presented in this manual.

1.2 SOCIAL VALUATION OF WATER SUPPLY PROJECTS

Costing is only one step in the process of valuing water supply projects from a social point of view, namely for the whole community (society) and not just for a single agent (the private investor). In turn, a social valuation, based on a comprehensive social cost-effectiveness or social cost-benefit analysis framework, is only one piece of the full set of information (economic, environmental, institutional, social, political, and so on) used by decision-makers in their decision-making process. Still, social valuation is an essential component of this decision-making process, and social costing an unavoidable element of social valuation. For more information on social cost-benefit analysis of drinking-water supply see: Cameron *et al.* (2011).

To understand the role and scope of our method, it is therefore important to set the costing of water supply projects in the general framework of social valuation of these projects. Figure 1.1 displays a flow chart – adapted from ADB (1999) – of the main steps in a full social cost-effectiveness or social cost-benefit analysis of a water supply project, showing the links with non-economic information (highlighted in grey). The components of the analysis covered by the method presented in this manual are highlighted in a dashed square in Figure 1.1.

Starting with a definition of the project rationale and objectives, a social valuation entails the following tasks:

- analysing and forecasting the need for project outputs;
- establishing the gap between future needs and the potential supply available with the optimized use of existing facilities;
- identifying options for meeting that gap in terms of technology, process, scale and location;
- performing least-cost analyses, using prices that assess the scarcity for the community of the marketed resources invested in a water supply project (social opportunity costs), in order to identify the best choices from a social perspective;
- assessing whether non-quantifiable benefits and quality considerations justify carrying out a least-cost analysis requiring a cost-effectiveness or a cost-benefit analysis framework;
- assessing, by means of a financial (liquidity) analysis, whether the project will be sustainable throughout its lifetime on the basis of cost-recovery, tariffs or subsidy (if any);
- testing for risks associated with the project through sensitivity and risk analyses;
- identifying and assessing the distributional effects of the project and its impact on poverty reduction.

In the context of this process of social valuation of water supply projects, we have developed a method for identifying and costing the technological options for producing the required water supply. This information allows for a least-cost analysis, to identify the best choices – from a social point of view – to bridge the gap between the identified needs and the supply that can be provided through the optimum use of existing facilities.

1.3 COSTING METHOD

The costing method presented in this manual is designed for use in low- and middle-income countries. It describes how to identify the cost components and collect the data needed to estimate the life cycle cost of the technical options that are feasible in the local context for providing access to safe drinking-water. This makes it possible to compare the costs of these options with a view to strengthening the basis for decision-making. The method can be applied either at local or at country level.

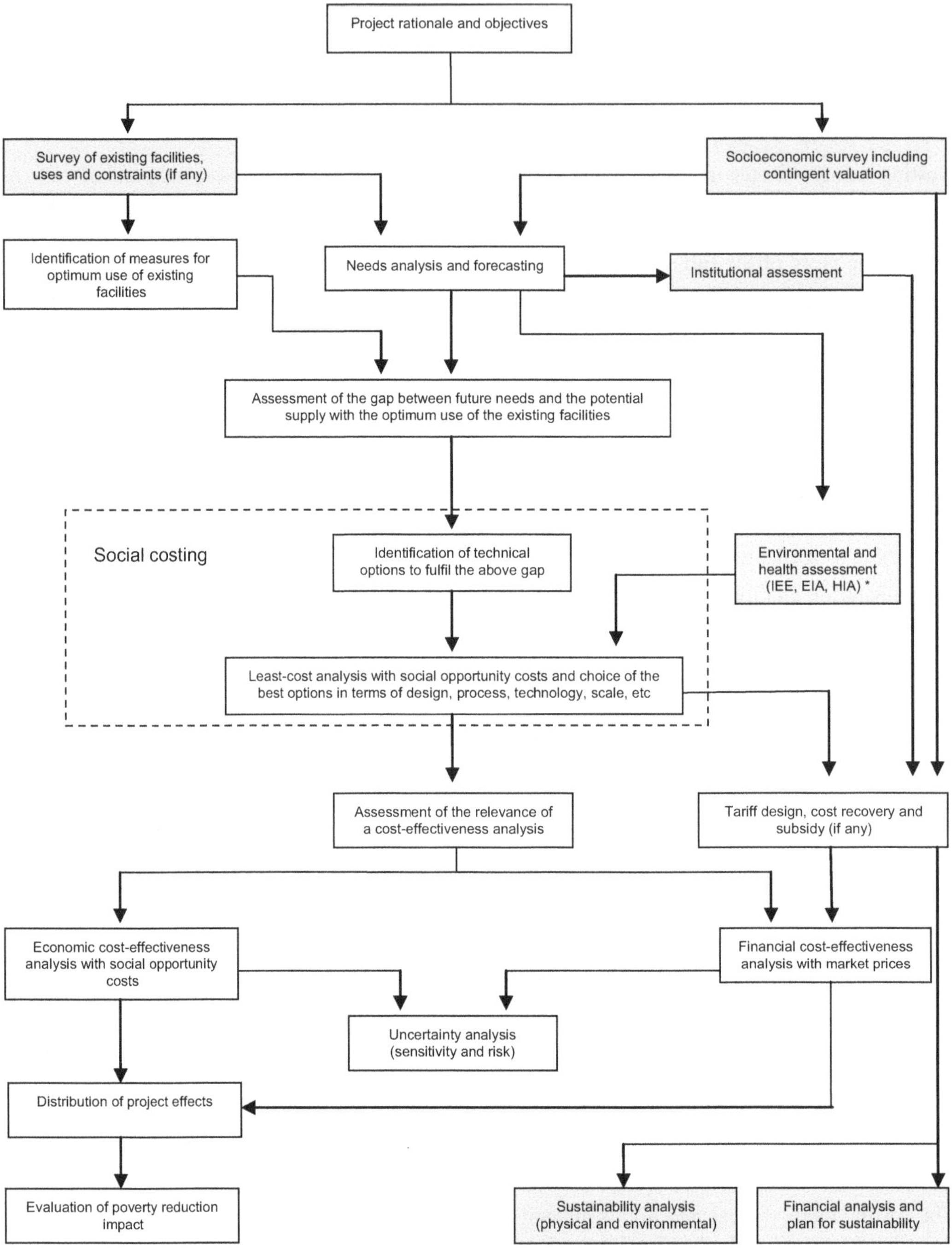

* IEE: initial environment examination; EIA: environmental impact assessment; HIA: health impact assessment.

Figure 1.1 Flow chart of a social valuation of a water supply project.

This manual provides practical guidance to facilitate and standardize the implementation of our costing method. It explains the process for costing 'improved' drinking-water supply technologies defined as technologies supplying drinking-water from an improved source. The concept of an improved drinking-water source has been defined by the WHO/UNICEF Joint Monitoring Programme for Water Supply and Sanitation (WHO/UNICEF, 2010) as one that, by the nature of its construction, adequately protects the source from outside contamination, in particular with faecal matter. As mentioned in section 1.1.1, the JMP definitions of improved sources of drinking-water include technologies bringing water either at a supply point shared by several households or at a supply point accessible only to single households. Therefore, our costing method of these technologies can be performed either at a community- or at a household-level, as suggested by Cumming *et al.* (2014).

To support implementation of the method, we have designed questionnaires and procedures for identifying and collecting data to evaluate these costs. We have also designed procedures for processing, aggregating and interpreting these data that will help identify least-cost options and contribute the cost components of a least-cost analysis, the aim being to promote the most efficient allocation of resources in terms of cost-benefit.

Along with the guidance provided in this manual, we have developed a basic tool to assist in applying the various procedures associated with costing a water supply system (please see page x for the download links). This is an Excel spreadsheet, which we call the Water Supply Costing Processor (WSCP). The spreadsheet offers a user-friendly way of identifying, collecting and processing the quantitative information needed to assess actual water supply projects. Once the required data are entered, the WSCP automatically provides both the detailed and the consolidated cost figures that are needed by decision-makers.

In this manual, the presentation of the costing method is organized as follows.

Chapter 2 gives an overview of the conceptual framework underlying our costing method.

Chapter 3 provides an overview of improved water supply technologies likely to deliver to final users' drinking-water collected form a source protected from outside contamination. More detailed descriptions of individual technologies are presented in Annex I.

Chapter 4 outlines the constraints that should be taken into account in selecting, from the available improved technologies, those suited for use in a specific setting. We refer to these technologies as being 'locally appropriate' for the project under assessment. Further details are given in Annex II. The suitability of a technology for a particular context is obviously of major importance, and in this manual we assume that our costing method will be applied only to appropriate technologies.

Chapter 5 presents the rationale of our costing process. We take an analytical approach that disaggregates an improved water supply technology into its essential components, namely, water collection, conveyance, storage, treatment and distribution. We have developed the questionnaires in Annexes III, IV and V to identify the main marketed resources invested in a water supply project. The questionnaires provide a way of collecting, for each component of a technology, four types of costs: investment costs; operational costs; maintenance costs; and other relevant costs, such as administration. These costs are evaluated using prices representing the social opportunity costs of the resources invested in the water supply project, regardless of who incurs the costs. Comparability between these different cost elements is achieved by discounting expenditures at different times to the same reference time. The picture of costs is then consolidated. Indicators of full and of unit costs can then be derived from this aggregated information, allowing the least-cost option to be chosen from among the various appropriate technologies assessed.

Chapter 6 shows how to apply our costing method to an actual project using an Excel spreadsheet (the Water Supply Costing Processor – WSCP). The use of this tool is illustrated with information from a field test in Peru presented in detail in Annex VI. The WSCP is a user-friendly tool for the collection and processing of the relevant information to assess a specific project. The ultimate aim of this tool and of the guidance provided in this manual is to present a standard methodology, which, if widely adopted, could provide a basis for cost comparison between projects as well as a standard basis for estimating and minimizing cost on individual projects.

To assess the scope and limits of the proposed costing method in a real setting, a series of field tests were designed and performed by local practitioners in selected countries. A first testing and assessment of the method reported in detail in Annex VI was commissioned from Teresa Lampoglia (2007), a local consultant from Peru. A second large scale testing opportunity was provided by a series of capacity building activities, reported in WHO (2008a) and WHO (2008b), organized by the WHO Water, Sanitation, Hygiene and Health programme in the Department of Public Health and Environment on 'Costing of Improved Drinking-water Systems for Low-income Communities', with participants from Cambodia, Indonesia, Lao PDR, Philippines, Thailand and Viet Nam. These tests were primarily intended to identify practical issues (limitations, difficulties, adaptability and user-friendliness) in the application of the manual and the use of the WSCP.

The results provided practical recommendations that have been implemented in the current version of the manual and the WSCP (notably implementation of the methodology presented in Chapter 5 for designing scenarios of the project life cycle production and a glossary of technical terms used in the WSCP). Field testing also identified the following issues, which deserve special attention in application of the method.

- Although the local conditions of a water supply project often provide overriding arguments in favour of a single technological option, within that option an insightful least-cost analysis can be conducted to determine how the different technology components can be implemented according to different economic alternatives.
- Successful implementation of the method requires a multi-disciplinary team and the creation of a partnership between sanitary engineers and economists.
- To support widespread utilisation of the method, it is important to complement the manual and the WSCP with a database of real-life case studies to present reliable estimates of both investment and recurrent costs and to illustrate justified choices of the social opportunity costs of resources and of social discount rate as defined in Chapter 5.
- An expansion of the costing methodology to non-market costs and benefits would be suitable for assessing water supply projects from a sustainable development perspective. This calls for an extension of the costing method to a more comprehensive social cost-effectiveness or social cost-benefit analysis framework. Research is continuing in this direction.

1.4 TARGET AUDIENCE

The manual can be used by professionals and practitioners from:

- national provincial and municipal drinking-water supply authorities;
- ministries of planning and of economic development;
- planning departments in ministries dealing with drinking-water supply (health, rural development, public works, water resources, others as relevant);
- national and international NGOs working in drinking-water supply;
- regional development banks and bilateral aid agencies;
- universities and engineering schools (departments of public health, health economics, civil engineering).

Chapter 2

Conceptual framework

2.1 IDENTIFYING LOCALLY APPROPRIATE TECHNOLOGIES

The first step in selecting a locally appropriate technology to provide a drinking-water supply to a low-income community is to obtain a comprehensive overview of the constraints facing that particular community in regard to its choice of an improved technology. This requires two surveys: a technical survey to assess existing water supply facilities and practices, as well as the local environmental conditions; and a socioeconomic survey to assess the social and economic conditions, including demographic profile, health status and institutional framework.

Based on this knowledge, and an understanding of the various options that exist for providing an improved technology of drinking-water, a selection can be made of drinking-water supply technologies that are appropriate for the particular community concerned. Once these locally appropriate technologies are identified, a least-cost analysis can be carried out to compare the options. Figure 2.1 displays a flow chart of the main steps of the method we suggest for selecting a least-cost and locally appropriate drinking-water supply technology.

2.2 COSTING A DRINKING-WATER SUPPLY TECHNOLOGY

The primary aim of a least-cost analysis is to identify least-cost options. The secondary aim is to generate the cost components of a cost-benefit analysis. The overarching purpose is to promote the most efficient allocation of resources: to select improved drinking-water technologies that achieve the maximum health benefits or welfare for a community at a minimum cost. In this context, the cost of selecting different technologies is understood to mean the social cost, namely the cost for the community at large.

Costs to households and utilities are based on the conditions of the locality, such as abundance or scarcity of water, land, labour, and so on. Such costs are also dependent on policy conditions determining discount and interest rates that reflect the opportunity cost of capital. Taxes, subsidies and donations should be ignored because they only reflect a transfer of money within the economy. Discrepancies between market and social costs may result from the government's equity goals and market imperfections. These discrepancies call for the use of 'shadow prices', namely prices prevailing in a competitive economy, assessing the cost of scarcity for the community of the marketed resources invested in a given water supply project.

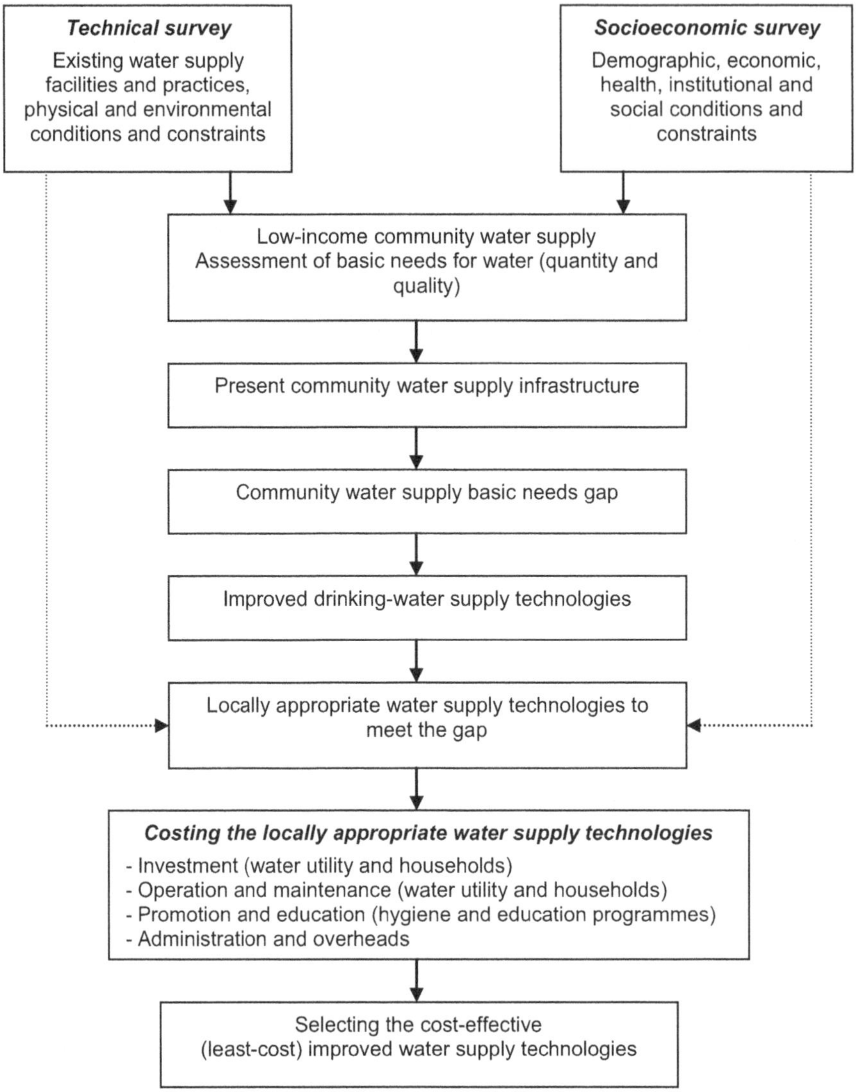

Figure 2.1 Approach for selecting a least-cost drinking-water supply technology.

Costing drinking-water technologies gives policy-makers an economic and financial baseline for their planning decisions. Each assessment presents the social opportunity cost of installing a particular technology.

The data required for a costing exercise comprise all the relevant costs. For a water supply project, this means all the related household and water utility costs. All of these costs must be included and evaluated. Any assumptions made must be consistent in regard to the different technologies.

Drinking-water supply technologies are valued in terms of the cost of construction, operation, maintenance, administration and overheads of the project. The opportunity costs of water (the benefits foregone in terms of environmental products and services, irrigation, livestock and aquaculture) have not been taken into consideration in the present version of our costing method.

Once the locally appropriate water supply technologies are identified, the next step is to estimate their life-cycle cost, in other words the cost of construction, operation, maintenance, administration and overheads. Assessing the entire life-cycle cost requires close cooperation between the economist, the planner and the engineer.

Other types of cost, such as those incurred in community organization, hygiene education and technical assistance, are not directly related to construction. They are normally considered to be complementary costs in a water supply project, and should therefore be quantified with regard to the selected technology.

Clearly, hygiene and education programmes add great value to the implementation of a chosen technology. Such programmes aim to change behavioural patterns, with a view to improving health in the community, and to improving living conditions in an environment free of disease. The concept of these programmes is not to be prescriptive, but rather to explain to people why they need to take certain actions and how such actions will help them.

Once the costs of locally appropriate water supply technologies have been estimated, the least-cost analysis makes it possible to select, from among those technologies, the best option from a social (community) perspective.

The conceptual framework for selecting least-cost drinking-water supply technologies generally deals with the ranking of mutually exclusive ways of producing the same output (for example drinking-water supply measured in m^3/day). The cost indicators to use in this least-cost analysis are presented in Chapter 5 of this manual.

Chapter 3

Improved drinking-water supply technologies for low-income communities

3.1 DRINKING-WATER SUPPLY SYSTEMS

The essential components of a drinking-water supply system in a low-income community are shown in Figure 3.1 and described below.

- *Water source*
  Any fresh water produced by the hydrological cycle, namely through evaporation of sea or lake water, condensation in the atmosphere and precipitation on ground as rain or snow, may be suitable as a drinking-water source. The main types of fresh water sources are:
  - *surface water*: stream, river, estuary or lake;
  - *groundwater*: springs, infiltration galleries, shallow wells or deep wells;
  - *rainwater*: direct precipitation catchments.

  Groundwater and surface water are the most important sources of drinking-water. In certain parts of the world, rainwater can also be a safe and plentiful source. The main considerations in selecting a drinking-water source are quantity and quality of water, and cost.

  The quantity of water available throughout the year should meet all the requirements of the community. The quality of water should be such that the treatment requirements are minimal. The cost involved in development, operation and maintenance of the water supply system should, within reason, be affordable to the consumers.
- *Water collection*
  An intake with suction facilities is required if the water is to be collected from a surface water source. A protected dug well is the more common collection device for the use of groundwater. A permanent roof is needed for the collection of rainwater.
- *Water conveyance*
  Water is normally conveyed by gravity or pumping. Dedicated structures carry the water from the water source to a storage tank or reservoir before treatment or water consumption.
- *Water storage*
  Reservoirs should have sufficient storage capacity to meet the anticipated water demand. Water is stored in reservoirs before treatment and distribution.

- *Water treatment*
  The type and degree of treatment required depends on the quality of water from the source. Groundwater is relatively free from disease bacteria but may be rich in mineral substances that may have to be removed. Surface water requires removal of turbidity and microorganisms. Water treatment commonly includes sedimentation, aeration, filtration, demineralization and disinfection.
- *Water distribution*
  The means of delivery of water to individual consumers varies. Water may be piped or carried in containers by various means of transport, or by household members or water vendors or others. Where a community water supply is available from standpipes, and there is no household distribution network, the standpipes should be located to minimize the laborious job of carrying water from the standpipe to the household.

Annex I presents the wide range of existing drinking-water supply technologies in detail.

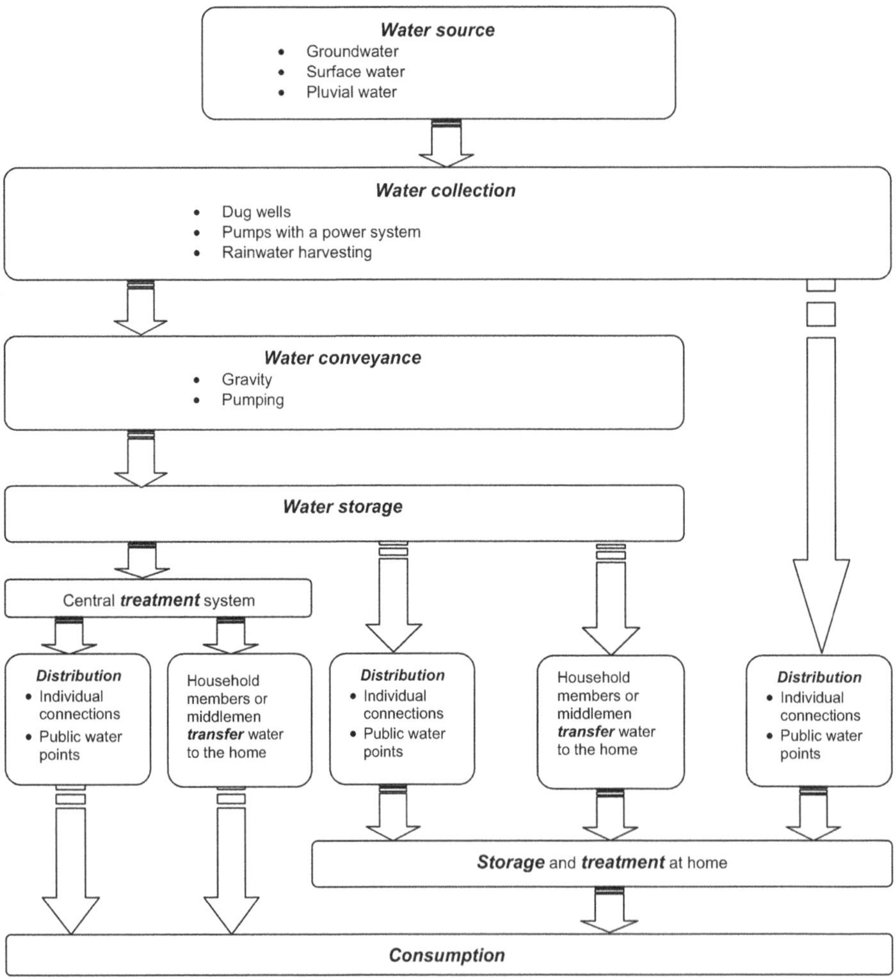

Figure 3.1 Components of a drinking-water supply process.

3.2 OBJECTIVES OF DRINKING-WATER SUPPLY

The main underlying objectives in any drinking-water supply system are described below.

- *Providing access to safe drinking-water*
  Safe water is free from microbial contaminants and toxic pollutants, in line with WHO's guideline values (WHO, 2008a). The WHO guidelines propose a risk assessment and management procedure that follows the generic outline provided by the Stockholm Framework (Bartram & Fewtrell, 2001). From a health standpoint, providing access to safe water reduces the risk of water-borne diseases.
- *Providing access to adequate quantities of domestic water*
  Water quantities should meet the requirements for effective sanitation and hygiene. From a health standpoint, providing access to adequate quantities of water reduces the risk of diseases that can be avoided by washing.
- *Making water easily available to consumers*
  The distance between the source of water and the point of consumption plays a critical role in water consumption. Privacy at the water point of use will also influence water consumption. To encourage the use of adequate water for personal and domestic consumption, sanitation and hygiene, it is important to improve access. From a health standpoint, distance from a water source relates to the risk of skeletal deformation and muscle damage from excessive carrying of water.

3.3 IMPROVED DRINKING-WATER SUPPLY TECHNOLOGIES

The water supply technologies we consider are those compatible with the MDG target of improving access to safe drinking-water in low-income communities. For operational purposes, the WHO/UNICEF Joint Monitoring Programme has defined drinking-water as the water used for normal domestic purposes, including consumption and hygiene, and has classified drinking-water sources as either 'improved' or 'unimproved' (WHO/UNICEF, 2010). Improved drinking-water sources are those that 'by the nature of their construction adequately protect the source from outside contamination, in particular with faecal matter'. Technologies that use improved sources, which we call improved drinking-water supply technologies, are more likely to secure a supply of safe drinking-water and, therefore, to promote the achievement of the MDG drinking-water target.

Improved drinking-water supply technologies are:

- piped water into dwelling, yard or plot
- public tap or standpipe
- tube well or borehole
- protected dug well
- protected spring
- rainwater collection

Unimproved drinking-water supply technologies are:

- unprotected dug well
- unprotected spring
- cart with small tank/drum
- tanker truck
- surface water (river, dam, lake, pond, stream, canal, irrigation channel)
- bottled water when the household uses water from an unimproved source for cooking and personal hygiene

An overview of improved drinking-water supply technologies illustrated by drawings (WHO/UNICEF, 2006) is presented below.

- *Piped water into dwelling, yard or plot*

Basically, piped water to the household is the most sophisticated technology. The drinking-water is treated and piped through house connections or yard taps. Drinking-water flows under pressure using a pump from the storage tank, satisfying the flow requirement demanded by the consumers, thereby achieving the three objectives mentioned in section 3.2.

- *Public tap or standpipe*

A public tap or standpipe provides water from a groundwater source. The water that people get from a public tap or standpipe is shared by more than one household. A public tap or standpipe requires a high level of water pressure. Generally, a public tap or standpipe is located in a public square, at a distance from the house. The average distance from user households influences water consumption, because of the laborious job of carrying of water.

- *Tube well or borehole*

Tube well or borehole technology is designed for the abstraction of groundwater (either at a shallow or at a deep level) using a pump operated in a suction mode. The suction pump draws water from a free or confined aquifer by creating a vacuum in the suction pipe. Shallow and medium-depth boreholes can be fitted with hand pumps, but deep boreholes will generally require a power-driven pump.

- *Protected dug well*

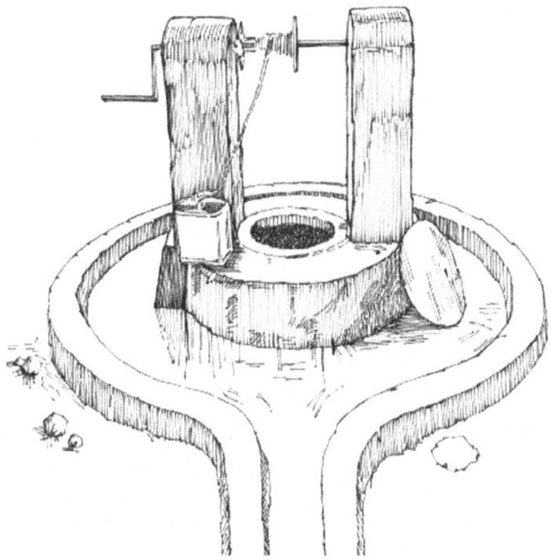

The dug well is a method of groundwater withdrawal in which a hole is dug in the ground to a depth below the groundwater level. Inflow into dug wells occurs as a result of the lowering of the water level in the well. Usually no special equipment or skills are required for the construction of a dug well. Protection is recommended to prevent bacterial contamination. The upper part of the well lining and the space between the wall and soil should be properly sealed has shown in the above illustration.

- *Protected spring*

Depending on the geographical and geological layout groundwater may be forced from underground to the surface. Generally, spring water emerges under the pressure of gravity or artesian gushing. With gravity, and depending on the water table, groundwater flows over an impervious layer onto the surface. With artesian gushing, groundwater flows upwards into the spring, and is forced under pressure to the surface. A protection chamber should be constructed around the spring, and the water should flow out of the chamber through a suitably placed pipe. The point at which the water emerges is a focus of pollution, thus digging a diversion ditch is highly recommended. Some spring sources may be highly polluted by soil organics, especially after heavy rains. In such cases, water requires local treatment, for example filtration and disinfection.

- *Rainwater collection*

In countries with considerable rainfall, rainwater is a potentially important source of water supply. Protected ponds replenished by rainwater are the main source of water supply in coastal areas. The approach of collecting, storing and using rainwater is highly feasible, but needs to be developed through adopting appropriate technologies.

The surface of roofs is used as a discharge. The rainwater is collected and stored in the dwelling, and is treated at home prior to use. The advantages of rainwater collection are simplicity and low cost. The disadvantages include the variability of precipitation and the resulting lack of guaranteed continuous service.

Although improved water supply technologies are necessary to provide safe drinking-water to low-income communities, they may not be sufficient if applied to drinking-water sources of inadequate quality. In areas where the air is polluted, rainwater may be contaminated by toxic aerosols. Dangerous levels of chemicals, such as naturally-occurring arsenic and fluoride, may be found in groundwater aquifers. Surface water from streams or rivers can carry infectious or toxic substances.

Therefore, before costing an improved drinking-water supply technology, the quality of the water at the source must be assessed and, if necessary, appropriate water treatment must be designed.

Chapter 4

Locally appropriate technologies

4.1 CONCEPT OF LOCALLY APPROPRIATE TECHNOLOGY

Only some of the technologies available for providing an improved drinking-water supply will in fact be suited to the specific setting of a project. Some technologies can be ruled out on the basis of local conditions (for example, arsenic levels in groundwater), because they will be incompatible with prevailing constraints or entail unacceptable risks. The constraints may relate specifically to local resources, or may arise from more general financial, economic, technical, environmental, socioeconomic, socio-cultural or institutional conditions. The risks relate to health.

Therefore, before tackling the costing of any technology, it is essential to identify local constraints and risks, and to discard all the technologies that are unable to overcome these constraints or that pose unacceptable risks.

The technologies that remain after this elimination process are referred to as 'locally appropriate' for the project under assessment and are those for which a least-cost analysis should be conducted.

4.2 CRITERIA FOR THE IDENTIFICATION OF CONSTRAINTS AND RISKS

It is useful to have a set of guiding criteria as a basis for identifying the constraints and risks facing a water supply project. The main screening criteria related to major constraints and risks are outlined below. This list is not exhaustive. Other criteria may be relevant, depending on the local conditions of the project being assessed.

4.2.1 Technical constraints

- *Distribution of the population*
  Rural settlements can be of two types.
  - Concentrated settlements where houses are contiguously located or next to each other, with a reasonably defined road plan.
  - Dispersed settlements where houses are isolated and distant from each other, with a considerable extension of unoccupied land and without a road plan.

    Non-rural (that is to say, urban or suburban) populations, by definition, inhabit concentrated settlements.

- *Minimum water availability*
  The quantity of water available depends not only on the characteristics of local water sources, but also on the knowledge of local sources. A suitable source will provide sufficient water during the highest demand days in dry periods, so that people are not obliged to turn to other sources of uncertain quality. This constraint is associated with local climatic conditions and weather patterns. It influences the replenishment of water sources and water use patterns. More water is used in warm and dry climates than in wet and cold climates. Normally, in temperate zones, water consumption on summer days is higher than the average daily demand on an annual basis.
- *Selection of the water source*
  Water should be drawn from the purest source available. This is especially important in low-and middle-income countries, where waterborne infectious diseases may be endemic, often against a backdrop of general ill health, and where acute health risks are much greater than in high-income countries. In high-income countries, chemical pollutants with longer term and chronic health effects present the greatest risk. Because of the epidemiological transition, many countries have to cope with both scenarios: a high incidence of acute, communicable diseases and of chronic, non-communicable diseases. In selecting the water source, the quantity of water that can be supplied and the accessibility of the source for potential users should also be taken into account.
- *Local resources*
  The availability of local resources, including the water source, is a determinant of the appropriateness of a water supply technology. Water is generally a location-specific resource and is normally a non-tradable output.

  Other local resources, such as local materials and land, are also important factors to be taken into account in identifying a locally appropriate water supply technology. In rural areas of low- and middle-income countries, labour costs are generally lower than capital costs. Therefore, preferred designs are labour intensive and rely on the use of local materials.
- *Location of the water source and water conveyance*
  Hydrological and topographical factors will determine the position and type of locally available water sources. Within these boundaries, the choice between gravity and pumping systems, as well as most other technical choices, primarily aim at minimizing cost.
- *Water quality and need for treatment*
  Because sources of water are usually known, the first step is to ascertain whether or not current supplies are of satisfactory quality. If not, it is essential to determine whether the poor quality results from natural circumstances (such as a high level of minerals or hazardous chemicals in groundwater) or from contamination of human origin.

  Generally, a superficial water source requires treatment before human consumption. The physicochemical characteristics of treated water should be in line with established quality norms in the existing legislation or with WHO guidelines values contained in WHO (2011). Therefore, the type of water treatment will depend on the quality of the water at the source.
- *Water use*
  The following factors may affect the use of a water supply system.
  - **Water acceptability**
    Safe drinking-water may be unacceptable because of its colour, turbidity or odour, or simply because it is different from the water the community is accustomed to.
  - **System capacity**
    The adequacy of the supply depends on all the design factors that established the required capacity in the service area, such as the design period, the population at the end of the design period, the

per capita average and peak water consumption, and the extension of the system to serve rural activities.

  - **Water distribution service**
    Many options are available but only solutions that adequately protect the water from outside contamination are recommended, although some may be more costly (see Annex I for details). Where the provision of residential metering is not foreseen, a layout that will allow district metering is desirable.
  - **Ownership of the water service**
    The rights to a water supply service do not usually belong to a single owner because they involve *water rights* and *land and property rights.*

    Traditional systems of community rights to water sources have developed in many countries where water is scarce. These water rights should be clearly defined. In low-income countries, however, these water rights are generally not constituted formally, and this could represent a technical as well as a legal constraint to the use of water sources.

    The ownership of the land where the water system is to be built will need to be established. Also, the ownership of the equipment and infrastructure will need to be clarified. Such ownership could be assigned to the government, to water utilities, to households, to the municipality, or to some other entity.
- *Sanitation*
  Where sanitation is to be provided, the choice of sanitation system will influence the amount of water required.

4.2.2 Environmental and social constraints

- *Environment*
  Water resources should be protected and conserved. The modalities of water sources development and management should be taken into account in an environmental assessment of water supply services. Water resources represent both an intrinsic value in terms of the biodiversity they support and an economic value in terms of tourism. Springs, rivers and lakes are often the focus of tourist areas.

  The flow from a water source varies seasonally, in both quantity and quality. Climate change may also affect the flow of water from a source.

  A water source is a scarce resource. Excessive use of a water source – generally for lack of water management – may cause environmental degradation. The risks to ecosystem services and products depend on location, climate, season, and potential use by people. Environmental services and products are closely linked to the level of biodiversity.

  Scarcity of water is relative, being determined by the type of technology applied, the intensity of application of the technology, and the quality and quantity of water. It is therefore important to evaluate the potential effects of water supply interventions in terms of environmental benefits and costs before constructing and operating water supply facilities.
- *Socioeconomic aspects*
  Collectively, socioeconomic aspects make up one of the most important factors to be considered in the selection of the type of water supply. They are measured as a function of the socioeconomic development level of a population. For example, empirical evidence shows that higher income populations have access to more complex water supply technologies than populations with lower living standards, for whom the technology needs to be simple.

To give an overview of which type of technology would be appropriate, a country's territory should be stratified by the poverty levels of its local population. The results should be shown in a quality of life map, using just a few categories of poverty (such as: very poor; poor; and not poor). The level of poverty of rural populations can be assessed using a method of standardized socioeconomic indicators, such as that suggested by FONCODES (1995) and used in Peru. The system of poverty indicators used for this purpose should include at least the following factors.

- **Education**
  Rate of illiteracy and rate of scholastic non-attendance.
- **Occupation**
  Number of children that work, and percentage of the economically active population without professional skills.
- **Housing conditions**
  Percentage of overcrowded households, and percentage of houses with precarious structures.
- **Basic services**
  Percentage of houses without a public water connection, percentage of houses without a public sewage network, and percentage of houses without electricity.
- **Health conditions of children**
  Rate of infant mortality, and percentage of chronically undernourished children.

These factors explain the situation of extreme poverty in a household. Other typologies could be applied (Deichmann, 1999). In a specific country context, the most important indicators should be tested for their relevance. Generally, in developing countries, the rural population presents levels of quality of life characterized as very poor and poor.

- *Financial aspects*
  The availability of financial resources for the project, whether drawn from local, national or international public sources, or from the private or the informal sector, must be adequate to meet all the project costs. These costs include planning, design, construction, promotion, operation and maintenance of the drinking-water supply, together with normal periodic expansion of the system.

  Investment decisions need to take account of financial sustainability (cost recovery through water pricing, level of subsidies), viability (net profits, average incremental financial cost, net present value, borrowing rates, financial rate of return), timing of investments in the project cycle context, and other such aspects.
- *Economic aspects*
  Reliable forecasts of water demand based on economic determinants, including the price of water and the income distribution among households, are needed to predict the likely consumption pattern of users. In many situations, the price of water (reflected in tariffs) is set below the real cost of water to make drinking-water affordable to poor households. In these cases, the government generally subsidizes the deficits.

  Other economic constraints to consider are: the level of hidden (sunk) costs, which are generally higher in urban areas than in rural areas; economies or diseconomies of scale; the opportunity cost of the water; the degree of saleability of water, in particular whether domestic users are end-users, or whether non-domestic users are intermediate or end-users; and users' willingness to pay.
- *Socio-cultural aspects*
  Social perspectives support particular water supply technologies and should be assessed by a social analysis of the community. Attitudes to water, reflected in views about the taste and quality of water, or in the quantity of water used in domestic or livelihood activities, can reveal whether water is considered a public good and whether people are willing to pay for a new or an improved service.

In planning the implementation of a water supply project, it is essential to understand the local culture and its potential for adapting to change. Sometimes a community does not embody a single culture. Ethnicity and gender inequalities are constraints that make it difficult to decide how best to implement a water supply project.

The government or the local authority should lead the implementation of water supply projects in a participatory process with the community. This process should promote, validate, plan and support the water supply project.

- *Institutional aspects*

  The institutional aspects of a water supply project concern the involvement of the different partners, along with the policy and regulatory frameworks in which the project will exist. The partners may include the public sector (national, regional and local government), the private sector (non-profit, commercial) and the users.

  For water supply projects to run productively within the applicable policy and regulatory frameworks, functioning institutions must be in place to support the project. Such institutions should have responsibility for providing water to the community, or at least providing technical assistance to support the supply of water to the community.

  When the partners in a water supply project work together, they can create an enabling environment for sustainable development. Without effective institutional partners, and without institutional arrangements that guarantee coordination and information flow between them, it is difficult to proceed with the installation of water supply facilities.

  The factors to be considered in the assessment of an institutional framework include the opportunities for and constraints on the recruitment of suitable staff, the capacity for human resources development (skills required), human resources management (clear descriptions of job functions linked to fair performance appraisal), logistical support (timely procurement for activities) and standardization (at national level).

4.3 ASSESSMENT OF WATER SUPPLY NEEDS

In order to assess water supply needs it is useful to differentiate between *effective demand* and *present water supply*. The present water supply is the observed water consumption. But if the availability of water at the source imposes restrictions on consumers (for example, because of interrupted service or fluctuating water quality) or if consumers decide to use other secondary sources of water, there is an unrestricted demand equal to the present water supply augmented by the unsatisfied demand. It is this virtual quantity that we shall call effective demand. In this sense, the supply of water from a source may restrict the effective demand.

In economic terms, the effective demand for water of a community of consumers is the quantity of water demanded of a given quality at a specified price. To estimate effective demand, therefore, involves a detailed demand analysis that allows an assessment of the water supply quantity and quality to be provided, of the size and timing of investments, and of the economic benefits.

Effective demand considers the demand for water of a certain quality (chemical composition, taste and odour, water pressure, reliability of supply, accessibility and convenience). An improved water quality is strongly correlated with higher levels of health. The combination of quality factors produces potable water, and users are willing to pay a higher price for a service level that result in a higher quality of water (house connection rather than public standpipe).

Affordability for domestic users is, therefore, an important criterion in selecting a water supply technology. In general, domestic users with high incomes are able and willing to pay more for a better quality of water. In relative terms, however, the people with the highest incomes will devote a smaller percentage of their income to water than people with lower incomes.

In this manual, we assume that analysing effective demand means identifying the community's needs for water used for normal domestic purposes including consumption and hygiene (see section 3.3). The health benefits of an improved drinking-water supply technology are a consequence of an improvement in water quality. Nevertheless, in some rural communities, the benefits for health of a greater quantity of water for domestic use exceeds those of better quality of water, because greater availability of water leads to improved practices of personal and domestic hygiene (for example, washing of hands, food washing and cleaning of the house).

A reliable assessment of water supply needs is a prerequisite for determining the appropriate technology to use. Urban and rural users have different patterns of drinking-water use, and may have different needs for water of higher quality. The needs assessment for urban and rural communities will therefore differ. Where the targeted beneficiaries of the water supply project have incomes and living standards below the country-specific poverty line, and where there is a high incidence of waterborne disease because of the lack of good quality water, the demand of this group is likely to be categorized as a basic need.

In urban areas, users are normally charged for the water supplied to them. In rural areas, a formal water supply may not exist, and rural users generally do not pay for the water they consume.

In both urban and rural areas, the parameters of water demand are: water price; household income; population growth; accessibility to the water source; transport cost of water; availability of alternative sources; opportunity cost of water; availability of water service levels; present use of water at peak and non-peak periods; weather (rainfall, temperature); and water lost through leakage or poaching[1].

For urban areas, an attempt can be made to derive an estimate of the price and income elasticity of demand. For rural areas, this is more difficult to assess but some estimation methods can be used, for example contingent valuation[2].

Other benefits of meeting a community's needs for water supply include the time saved if water no longer has to be carried to the household. With more time available, women and men can increase their income-earning activities.

4.4 SELECTION OF LOCALLY APPROPRIATE WATER SUPPLY TECHNOLOGIES

In any drinking-water project, the appropriateness of potential technologies will need to be tested in terms of the constraints presented in section 4.2, faced by the project. In this regard, the upgrading of an existing facility should be treated as a 'new' technology and hence be subjected to similar testing.

The aim is to choose the locally appropriate water supply technology that will comply with the expectations and needs of the users. In selecting water supply technologies for consideration, all constraints and risks, as well as their interrelationships, should be examined. This will allow for an efficient choice of technology that will maximize the social welfare of the community concerned.

From the very beginning, the activities leading to the installation and operation of water supply systems should recognize the role of the beneficiaries.

Specially designed questionnaires should be used to collect and record data. Technical and social constraints apparent from the data collected will narrow down the choice of technology (see Annex I

[1]The difference between the water produced (and distributed) and the water sold, or the water produced but not sold.

[2]It is a direct method of non-market valuation in which users are asked directly about their willingness to pay for a specific hypothetical quantity or quality of goods or services such as water supply.

for descriptions of the most commonly used water supply technologies in low-income communities). Other constraints such as financial, environmental, socio-cultural and institutional) as well as health and environmental risks, should be evaluated in the local context.

The basic data to be obtained from technical and socioeconomic questionnaires are outlined in the box below. Model questionnaires are presented in Annex II.

OUTLINE OF A QUESTIONNAIRE ADDRESSING TECHNICAL ISSUES

General information

- Name of country
- Name and administrative location (state, province, region, district, municipality) of the area or sector
- Location of water supply project
- Sources of data

Settlement characteristics

- Description of the locality or area
- Demography
- Number and type of houses (density in the locality and in the project area)
- Distances to schools, postal services, public services, and infrastructure in the community

Hydrology and climate

- Type of hydrology of the area
- Type and quality of the soil
- Temperatures and periods with and without precipitation

Communication and transport

- Quality of the roads (availability and feasibility of transit in the winter or in the rainy season)
- Distances to populated areas
- Access to other means of communication

Uses of water

- Human consumption
- Hygiene
- Other domestic uses
- Irrigated agriculture
- Livestock consumption
- Energy
- Industry
- Recreation

Existing water and sanitation conditions

- Type of present drinking-water supply (source, infrastructure, maintenance, water quality)
- Disposal of excreta (type of facility, infrastructure, maintenance)

Electricity

- Responsible authority or company
- Supply capacity
- Reliability of supply
- Power installed and actually available in the locality

OUTLINE OF A QUESTIONNAIRE ADDRESSING SOCIAL ISSUES

General information

- Location and source of data
- Education levels and demographic structure

Prevailing labour conditions

- Levels of income and income distribution
- Sources of available work in the locality
- Employment levels and seasonal fluctuation

Social organization and activities

- Local organizations
- Attitude of the community to the sanitation
- Possible contributions of the community to the water supply project

The decision process displayed in Figure 4.1 shows the different phases in selecting appropriate water supply technologies for low-income communities. It may serve as a useful tool or reference checklist to help decision-makers identify and select the most appropriate water supply technology for use in rural areas, for which a least-cost analysis should be conducted. Although this selection process is applicable to most situations in low- and middle-income countries, there may occasionally be circumstances where the most appropriate technology is not the one suggested by the process.

The selection process is particularly useful when there is no formal water supply technology in the target community. In some cases, any existing water supply technology can influence the selection of the appropriate technology in ways that the process cannot fully capture.

It is important to consider existing or planned water supply technologies in neighbouring areas because the possibility of sharing these facilities may enable the community to reduce its costs through, for example, economies of scale.

After selecting the appropriate water supply technology, the following points need to be clarified to ensure that the process of costing the technology is valid.

- *Definition of a baseline*
  The baseline refers to the present situation with regard to existing water supply technologies in the community. These facilities may be improved or unimproved. In some cases, the community may lack access to improved sources, but a natural unimproved source exists.
- *Definition of the level of service to be considered*
  The level of service may be set, for example, as a piped house connection to each household of 7 or more persons, providing each household with a 24-hour water supply, sufficient for a per capita consumption of 100 litres per day.
- *Definition of the level of service to be considered*
  The level of service may be set, for example, as a piped house connection to each household of 7 or more persons, providing each household with a 24-hour water supply, sufficient for a per capita consumption of 100 litres per day.
- *Determination of the scope of an intervention*
  The scope of an intervention may be determined, for example, on the basis of the assumption that the existing network only requires some piping to serve a new home, or that the water intake,

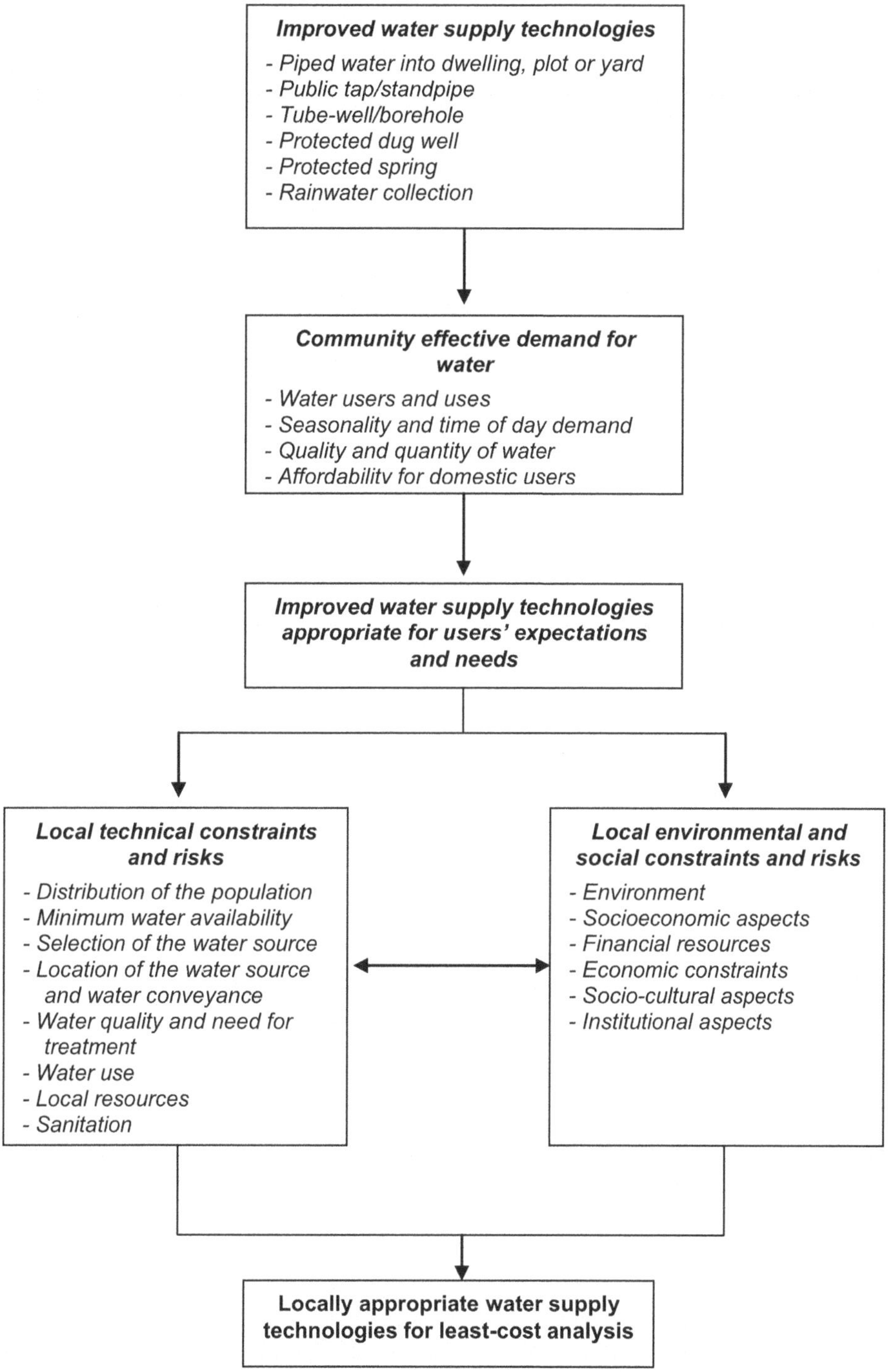

Figure 4.1 Selection of locally appropriate water supply technologies.

transmission and treatment works will all need to be expanded if a system of standpipes is to be converted to house connections.

- *Identification of the different costs to be included*
  In practice, the costs to be included are: water utility and household construction costs; water utility and household operation and maintenance costs; and water utility and community promotional, education and overhead costs.

Although the main factor in comparing alternatives is cost, other important financially non-quantifiable factors should also be taken into account. Such non-quantifiable factors include:

- *Water quality*
  A water source of higher quality is preferable to one of lower quality.
- *Availability of local materials*
  A system that does not depend upon spare parts or chemicals imported from abroad is to be preferred, even if costs appear somewhat greater.
- *Reliability*
  A gravity system is inherently more reliable and hence preferable to a pumped system because it operates without pump, fuel and electric power, and requires less maintenance and fewer skilled people.

Chapter 5
Costing method

5.1 RATIONALE AND ISSUES

The primary aim of this costing method is to put an economic value on the opportunity cost to the national economy of installing a drinking-water supply system or providing a drinking-water supply service. This economic value can be used in a more comprehensive social cost-benefit analysis (SCBA) that compares all the resources that are scarce in society, invested in a water supply project to the complete set of project outcomes contributing to improve the quality of life and health of project beneficiaries (Cameron *et al.* 2011).

Because a SCBA looks at the impacts of a project for the whole community, and not just for a single agent (the private investor), the social costs and benefits may differ significantly from those of a private economic evaluation of the same project.

Two reasons explain these differences in values:

- In a SCBA, the prices of marketed goods and services should be set to motivate their users to individually choose less expensive options, so as to generate less expensive solutions for the community. These prices (including interest rates, wage rates and foreign exchange rates) should be those prevailing in a competitive economy. Hence, when actual market prices involve significant distortions arising from market imperfections, they should be replaced by 'shadow prices' reflecting the scarcity cost, for the community, of the marketed resources invested in a social project.
- In a SCBA it is advisable to account also for the non-market costs and benefits of a project, which entail no explicit exchange between economic agents, such as those generated by the external effects or the use of public goods. Costing a water supply project from a social perspective will require, in particular, taking into account: the external costs arising out of environmental damage; the opportunity costs of the benefits forgone because of diverting raw water from productive activities, such as agriculture, to non-productive activities, for example basic domestic uses; and the depletion premiums valuing the cost of conserving water sources, such as groundwater aquifers (reservoirs).

Given the difficulty of obtaining acceptable values for such non-market costs, we decided to limit our costing method to *marketed costs* (the cost of the resources used in a water supply project that can be provided by markets), regardless of who (utility, households, government, or other) incurs them. Note that by limiting our analysis to marketed costs, our method may be used, with the required adaptations having been made, to perform a private economic costing of a water supply project (based on actual non-competitive market prices). This makes it possible to assess the financial requirements for implementing, operating and maintaining the project during its life-cycle.

The proposed method for costing appropriate improved water supply technologies relies on an analytic approach that disaggregates the water supply process into the main components described in section 3.1, namely: water collection; water conveyance; water storage; water treatment; and water distribution to final users. This method of costing a water supply system was inspired by the so-called activity-based costing (ABC) method. This ABC method of cost accounting breaks down the water supply process into a sequence of single activities, each measured using quantitative indicators that are strongly correlated with the cost of the activity. However, while the ABC method is suitable for costing complex water supply technologies, such as those used in modern urban communities, our method is better suited to tackle the costing of the simpler water supply technologies used by low-income communities in suburban and rural settings.

Our costing method requires three steps.

- Identifying the cost components of a water supply technology.
- Collecting data on the identified cost components.
- Analyzing the collected data on costs to derive cost indicators for the life-cycle of the project, thus making it possible to perform a comprehensive least-cost analysis of alternative water supply technologies.

Identifying and collecting the relevant cost components of a water supply system is far from trivial. Government decision-makers are naturally most keen to know what financial costs their departmental budgets will incur. But from an economic perspective, it is just as important to know about significant household costs (such as plumbing) or opportunity costs (such as the time spent collecting water) that a water utility will never see. Providing guidance on how to identify costs at an appropriately disaggregated level, and on where to find the relevant data to estimate the cost of a water supply technology, lies at the heart of this exercise.

The following points relating to our analytic costing method need emphasis.

- Clear criteria are needed as a basis for deciding which costs are to be included and which are to be excluded.
- The costing should be comprehensive, covering hardware costs (construction, operation and maintenance) as well as software costs (administration, training, promotion and education).
- The issue of non-marketed costs needs to be addressed carefully. In particular, community and household contributions to operation and maintenance should be realistically valued and not simply taken as being provided free of charge.
- The market prices of goods and services need to be adjusted if they do not reflect the national opportunity costs of resources.
- The costs of hygiene promotion and education that are part of a general health campaign should be included in the costs of water supply technologies, even though they may be incurred in another public sector than the one with core drinking-water supply responsibilities in its mandate.
- The additional costs of moving from one service level to a higher level (by improving, upgrading, or extending a planned water supply project) need to be calculated. For example, the total cost of building a system of standpipes and then extending the system to serve house connections is not the

same as the total cost of building a system designed, from the beginning, for house connections. We cannot assume that the cost of the extension is simply the difference between the cost of a system of house connections and the cost of a system of standpipes.

5.2 TYPOLOGY OF COSTS

For each main component of a drinking-water supply system listed in section 3.1, we consider four types of costs: investment; operation; maintenance; and other recurrent costs.

A brief description is given below. Annex III presents the details of investment, operation and maintenance costs for the six improved water supply technologies described in section 3.3.

5.2.1 Investment costs

The investment costs include costs of construction activities, equipment and materials, including the following costs.

- *Preliminary studies*
  Studies carried out at the pre-investment stage, to understand the technical, social and environmental aspects of the construction project.
- *Equipment*
  Capital cost of infrastructure and power systems.
- *Local materials*
  Materials produced in the country where the project is implemented.
- *Imported materials*
  Materials of foreign origin.
- *Skilled workers*
  Qualified workers including specialist civil engineers, civil constructors, technical staff, and social science professionals.
- *Unskilled workers*
  Manual workers working, for example, in excavation, in carrying of materials, or in cleaning. In many instances, community members provide the workforce for these activities.
- *Semi-skilled workers*
  Workers somewhere between skilled and unskilled, depending on the type of work.
- *Other investment costs*
  Management, administration, coordination, logistics, transport, communications and office costs, as well as outsource tasks and quality control costs, along with any other unassigned investment costs.
- *Contingencies*
  Incidental investment costs to allow for adverse conditions that might increase expected costs, usually estimated as a lump-sum amount or a percentage of the total investment costs.

Investment costs in each identified component of a water supply system (collection, conveyance, treatment, storage and distribution) are presented in Figure 5.1. Note that the inputs to be assessed are categorized as materials (M), equipment (E) or labour (L).

5.2.2 Operation and maintenance costs

Operation and maintenance are recurrent activities required to keep a water supply system in operation and in good condition during its life-cycle.

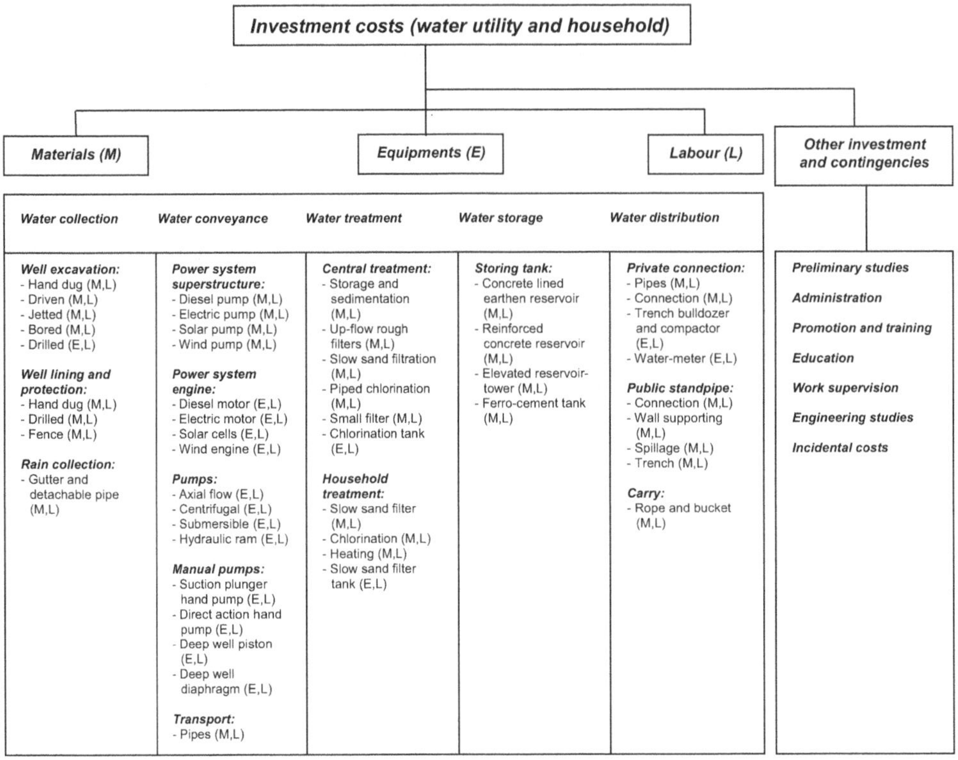

Figure 5.1 Investment costs.

Operation costs include expenses for personnel, chemicals, electricity, fuel, materials, office supplies and building rents. In general, operation costs are constant over time if the prices of inputs (for example \$/KWh) and the activity level or output (for example m^3 of drinking-water delivered) remain the same. In this case, operation costs can be assessed as a constant annuity over the life-cycle of the project. In the opposite case (for example with a considerable rise in energy prices or a gradual increase of the level of services provided by the water supply facility), a scenario should be designed for all those components of the operation costs that are expected to change during the water supply system life-cycle.

Operation costs in each identified component of a water supply system (collection, conveyance, treatment, storage and distribution) are presented in Figure 5.2. Note that the inputs to be assessed are categorized as materials (M), power services (PS) or labour (L).

Maintenance costs include all the expenses for maintaining and repairing infrastructure, equipment and vehicles. In general, maintenance costs of equipment increase over time and they depend on the activity level (ideally, there should be 100% use of the available equipment). Therefore, to assess these costs it is important to have the maintenance plan for the equipment, as formulated by the provider. The planning of maintenance should identify all the activities involved in maintenance and the activity levels for implementation (hours of work by activity, replacement parts, repairs procedures, inputs, and so on) and the timing or frequency at which these activities should be performed. If these activities and activity levels evolve over time, a scenario should be designed for all those components of the maintenance costs that are expected to change during the system's life-cycle.

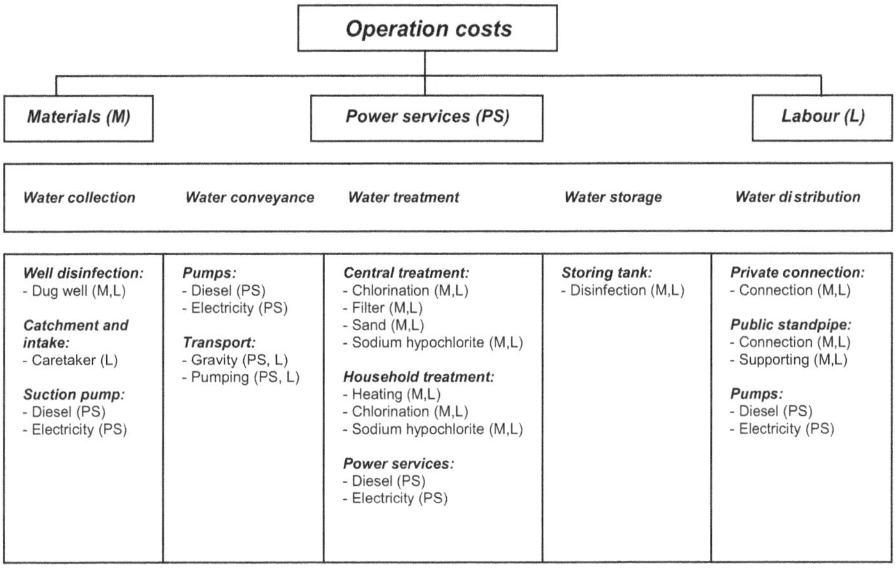

Figure 5.2 Operation costs.

Maintenance costs in each identified activity of water supply (collection, conveyance, treatment, storage and distribution) are presented in Figure 5.3. Note that the inputs to be assessed are categorized as materials (M), equipment (E) or labour (L).

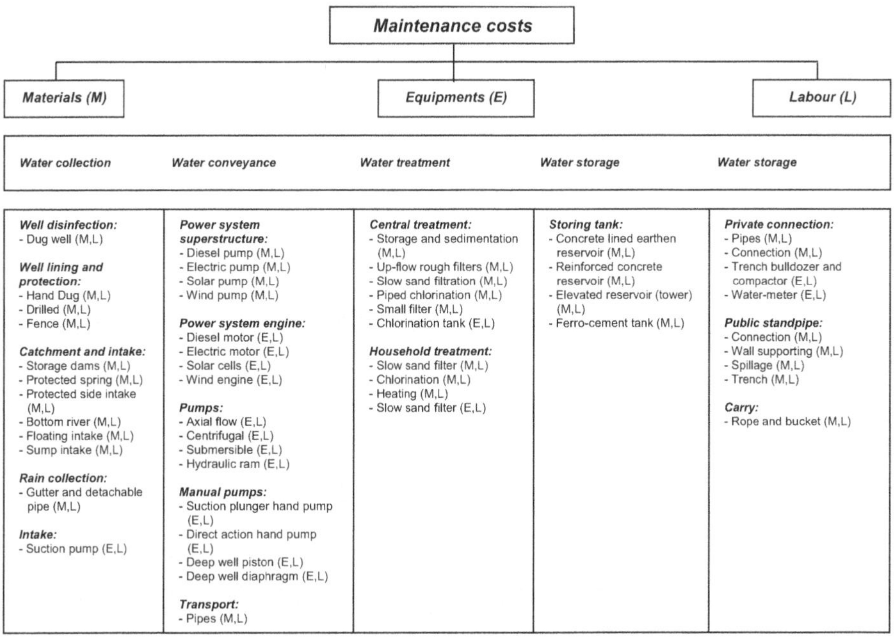

Figure 5.3 Maintenance costs.

5.2.3 Other recurrent costs

These costs encompass the operational costs of a water supply technology, reflecting the correct functioning of the system. In this context, the most important costs are the administrative costs of the system, as well as the relevant training, promotional and educational costs. The quantification of these costs depends on the design parameters of each water supply technology. For simplicity, they are therefore commonly estimated as a percentage of the construction costs.

These costs may be explained as follows.

- *Administrative costs*
  Costs related to the management and administration of water supply services. They are usually estimated as a percentage of the construction cost.
- *Costs of training in administration, operation and maintenance*
  Costs of training activities and programmes necessary to develop managerial and technical skills of local managers of water supply services and their staff. These training activities should develop capacities for dealing with accounting, budgeting, inventory control, personnel management and public relations. Usually, the municipality or the utility cover these training costs. Like administrative costs, training costs are usually estimated as a percentage of the construction cost.
- *Promotional and educational costs*
  Costs of promotion and education related to health and hygiene taking into account all specific activities to improve the hygiene habits of rural communities. These costs include the fees for trainers, the cost of services, the cost of acquiring or renting equipment, and the costs of materials and inputs, transport and logistical support. For example, in rural localities in Peru, the average time required for promotional interventions is one year and the estimated costs are 10–15% of the construction cost.

Figure 5.4 shows the main components of these costs.

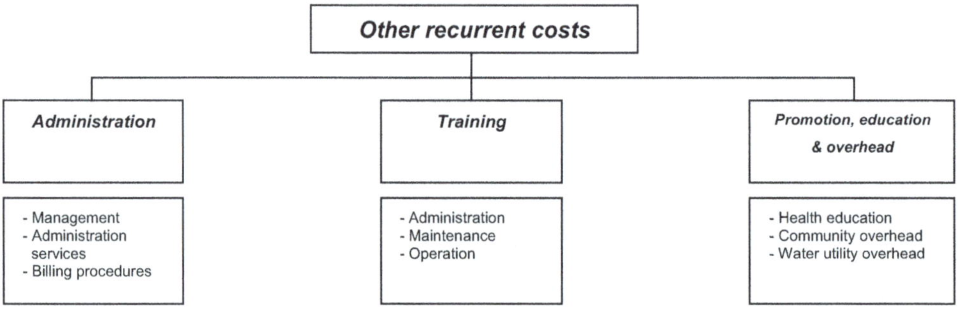

Figure 5.4 Other recurrent costs.

5.3 SOURCES OF DATA ON COSTS

Identifying sources of the data needed to estimate costs depends on the type of cost. Some potential data sources for different types of cost are listed below.

- *Construction costs*
  Information sources for these costs are the easiest to identify, as they can usually be determined using bid contracts. For small-scale works that are not under such contracts, other sources of data must be identified, for example nongovernmental organizations.

- *Operation and maintenance costs*
  Sources for information on these costs are more problematic to identify than those for construction costs, as they require someone to manage cost records effectively over the years. Information may be available from the utilities in larger towns and cities, and from local entities in rural settings.
- *Promotional costs*
  These are best determined using those involved in promotional programmes as the source, for example government agencies and nongovernmental organizations.

The practitioner (a sanitary engineer is the appropriate person to perform these tasks) should identify the sources of data among local entities and institutions. As a practical recommendation, we suggest consulting the following entities.

- A local authority such as a committee, a directorate or a similar entity, that administers the water supply infrastructure for rural communities. The local authority should at least have information about the expenses related to the construction, maintenance and operation of water supply systems.
- A municipal, council or city office with a public works department in charge of the local infrastructure of any water supply programme. Such an office should have data on the water supply technologies that they administer.
- A public works secretary, a ministry of public works or a government institution in charge of the planning of construction and the control of water supply technologies in rural communities.
- Generally, these agencies have studies of engineering work that have been audited by a multilateral organization, such as the World Bank, the Inter-American Development Bank, the Asian Development Bank or the African Development Bank.
- With regard to expenses in promotion and hygiene education, the practitioner should consult these institutions and the Health Secretary or Ministry of Health of the country.
- Other sources, when the public sources cannot provide the necessary data. These other sources could be private sources such as construction firms, or consulting firms, or anyone who has participated in the construction of water supply facilities or in preliminary studies related to water supply technologies in rural communities.

5.4 COSTING QUESTIONNAIRES

To facilitate the identification and collection of information about all these costs, a set of costing questionnaires have been designed describing the main marketed resources (materials, equipment, labour, power services, and so on) invested in the project. These questionnaires are presented in Annex IV. Based on the tabulation of the costs of improved water supply technologies presented in Annex III, each costing questionnaire describes the use of a given resource within one of the four categories of costs described above (investment, operation, maintenance and other recurrent costs), by detailing the use of the resource for each main activity of a water supply technology (collection, conveyance, storage, treatment, distribution). The descriptions of resource use are provided for three levels of aggregation.

- *Item level*
  Aggregation level corresponding to the technologies that can be used to perform a single activity of an improved water supply system; for example, water collection can be carried out using different technologies, depending on the water source (such as a gutter for rainwater or a well for groundwater).

- *Sub-item level*
  Aggregation level corresponding to the particular technical device that can be used in practice to implement a technology; for example, a wood gutter, a galvanized gutter or a PVC gutter can be used to collect rainwater.
- *Input level*
  Aggregation level corresponding to a breakdown of the sub-item description of a technical device, at a more detailed level at which cost data can be collected; for example, a wood gutter structure and nails would be the (input level) local materials necessary to install a wood gutter for collecting rainwater.

Once the resources involved in the realization of a water supply project have been identified, the next step consists in looking for the data sources needed to quantify the resources invested in the water supply project. Depending on the data sources available, this quantification can be performed at a disaggregated level (input level breakdown) or at a more aggregated level (sub-item or item breakdown).

- Use of the disaggregated level option requires the physical measurement of the invested resources described in the input level breakdown. This level of disaggregation is recommended to provide clarity in the process of valuing economic resources in monetary terms, and it may enable the costing to be used in other settings.
- The sub-item or item aggregation level options have been devised for situations where historical or bid data sources lack sufficiently detailed information to allow for costing at a disaggregated level. In such cases, only the monetary value of a cost component at a sub-item or item level is required.

5.5 SOCIAL COSTING OF WATER SUPPLY PROJECTS

Once the locally appropriate improved technologies have been identified by the project engineer, it is necessary to rank feasible projects according to some meaningful criteria in order to select those technologies most appropriate for the planned project.

If the identified technologies are mutually exclusive and provide the same outcomes, they can be compared simply on the basis of cost. The least-cost alternative can then be selected as the economically optimal solution.

When the appropriate water supply technologies offer different levels or quality of service, a least-cost choice will not necessarily be the economically optimal solution because some technologies have benefits that compensate for their higher costs. This is the most common situation. Costing analysis will not provide sufficient information to select the most appropriate technology, but will still contribute an important input to the process of assessing water supply projects from a more comprehensive socioeconomic point of view. Indeed, when there are differences in outcome, a cost comparison of locally appropriate technologies, if properly conducted, can reveal the trade-off costs corresponding to the different levels or quality of service provided by the competing technologies. This is useful information for the decision-maker in deciding how much to pay to obtain various standards of service. In this sense, costing should be seen as a further screening of the various water supply technologies that have already passed the test of technical and social feasibility.

5.5.1 Principles of social costing

The primary intent of social costing is to develop a value of the opportunity cost to the national economy of providing a given water supply service. In practice, there are three principles to follow in preparing estimates of the social costs of a water supply project:

- All relevant costs to the economy, regardless of who incurs them, must be included.
- Each cost must be properly evaluated using competitive market prices representing the national opportunity costs of the resources invested in the water supply project.
- The assumptions used for costing different technologies must be mutually consistent and comparable to allow least-cost analyses.

5.5.2 Inclusion of all relevant costs

All costs to the economy should be included, irrespective of whether they are borne by the utility, the households, the national government and/or other entities. Some financial costs, such as taxes and subsidies, should be ignored because these represent only a transfer of money within the economy and not a cost to it.

The determination of which costs to include should be based on a comparison, over time, between the situation with the project and the situation without the project. In particular, it is essential to estimate how the current situation would improve or deteriorate over the project period without implementation of the project.

In addition, the boundaries of the water supply project must be viewed from a broad perspective so that all relevant costs will be included.

5.5.3 Estimating competitive market prices of resources

In many instances, the socioeconomic policies of governments are only loosely related to purely economic objectives. Therefore, actual market prices do not always reflect real resource scarcities within the national economy, because they may differ from the prices generated by competitive markets. Such distorted prices should be converted into virtual competitive market prices by using 'shadow factors' that express the ratio of an economically efficient or 'shadow' price to its actual (economically inefficient) market price.

In most developing economies, such price distortions are endemic. In particular, they affect the wage rates of unskilled labour, because of substantial 'structural unemployment', and foreign exchange rates, because of barriers to free trade that governments set through quotas, taxes and subsidies.

Estimating these conversion factors is a difficult and subjective task. It requires intimate knowledge of a country's economy, as well as the expertise of the economists in charge of macroeconomic policy and the planning of social projects. Good sources of these figures include the Ministry of Economic Development or the Ministry of Planning of the country concerned. The World Bank, the Inter-American Development Bank, the Asian Development Bank or the African Development Bank, are also potential sources of estimates for the shadow factors. Young and Haveman (1985) provide an introduction to this issue of shadow pricing.

In the economic costing of water supply technologies, three kinds of shadow factors must be incorporated into the analysis.

- *Unskilled labour wage*

  Because of minimum wage legislation, unskilled labour is economically overvalued, that is, the wage is higher than the real value of labour productivity, resulting in unemployment. This means that, if there is widespread unemployment among the unskilled labour force in a country, the shadow factor for unskilled labour would be close to zero. This is because there is almost no cost to the national economy resulting from the employment of such workers, since they would be otherwise unemployed.

  In contrast, if a country has few unemployed unskilled workers, then the shadow factor would be 1, as this situation is an indicator that the market wage for such labour fairly reflects the economic scarcity of this resource. In developing countries, the shadow factor for unskilled labour is between 0.5 and 1.

- *Rate of exchange for foreign currency*
  Many governments do not permit free movement of their national currency in the international foreign exchange markets. They prefer to fix the value of their country's currency in relation to that of their major trading partner. As a result, the price for imported materials and equipment expressed in the national currency is below its international market value. The same result may also be achieved through a system of import restrictions.
  The shadow factor for the rate of exchange for a foreign currency to be used in such situations, namely the ratio of the shadow rate of exchange to the official rate of exchange fixed by the government, will be greater than 1.
- *Price of land, water, power and other direct inputs*
  The prices of some inputs to water supply systems are controlled by governments or include government subsidies. For an economic costing of water supply systems, these prices should be calculated as if the inputs where sold on competitive markets.
  For land, a good approximation of its shadow price can be obtained by reviewing the records of recent sales of similar land in the area. For water and power, it is not possible to estimate directly what a free market price would be when the government has a monopoly in their production. Nevertheless, an approximation of the shadow prices of these inputs can be obtained by computing their average incremental cost of production, according to the formula presented in the next section.

In addition to these adjustments to the distorted actual market prices of resources, a social costing of water supply systems must value the physical capital components of such systems at their replacement costs rather than at their historical costs, because historical costs should be disregarded in making decisions about future investments. This means that the costing of physical capital components should be based on bids collected by tender.

5.5.4 Computing cost indicators for least-cost analyses

Applying the above-mentioned costing principles requires setting up a time line of expenditures to implement, operate and maintain the water supply technology throughout its life-cycle. All water-supply technologies involve investment, operation and maintenance costs. These costs occur at different points in time, some annually, like operational costs, others less regularly, like maintenance and replacement costs, and some only once, like the installation of heavy infrastructure.

To perform least-cost analyses of alternative technologies that provide drinking-water services during their design lifetimes, it is necessary to consolidate these time sequences of expenditures into a full-cost total for each project. This is done by computing the present value (at a specified reference date) of these expenditures, using a discount rate that is appropriate for social projects.

More precisely, by assuming a design lifetime of the project of T years, and denoting by C_t the total costs incurred in year t of the project life-cycle and by i the annual social discount rate, the *full cost present value* (*FCPV*) of the project is computed using the following formula:

$$FCPV = \sum_{t=1}^{T} \frac{C_t}{(1+i)^{t-1}}.$$

This formula implicitly assumes that annual costs all occur at the beginning of the year.

The choice of the social discount rate is a matter of considerable debate (Zhuang *et al.* 2007; Pannell & Schilizzi, 2008). In an ideal competitive economy without market imperfections, and with a perfect financial market where any economic agent can lend and borrow any amount of money at a unique rate of interest, the social (as well as the private) discount rate is equal to the rate of interest. This is because, for an economic agent, it is equivalent to hold two different cash-flow time sequences having the same discounted value at the current interest rate, as it can convert (without costs) any of these sequences into the other by using the financial market. This unique rate also equals both the rate of time preference and the opportunity cost of capital.

- The *rate of time preference* (RTP) is the rate of return on loans, which motivates a consumer to save by postponing a marginal unit of current consumption in exchange for more future consumption.
- The *opportunity cost of capital* (OCC) is the rate of payment on loans, which motivates an investor to borrow a marginal unit of capital to fund a productive activity that generates a higher future return.

Market imperfections unbalance this equality by creating a gap between RTP and OCC (with the former generally lower than the latter), and make both differ from the market rate of interest. In such circumstances, what rate should be used to discount future benefits and costs in social cost-benefit analysis? There are two main contenders.

- The use of the *social rate of time preference* (SRTP) has been advocated on the grounds that, unlike a private discount rate, the social discount rate should not merely express the average cost of capital invested in social projects but the inter-temporal substitution rate in consumption used to trade off the level of present national consumption against that of investments increasing future consumption. This social discount rate can be revealed by examining a development plan for the national economy. By means of a macro-econometric growth model, it is possible to simulate the impact of policies aiming at marginally increasing investments during one year (by decreasing consumption), to foster future production and consequently future consumption.
- The use of the *social opportunity cost of capital* (SOCC) has been suggested for those situations where public and private sectors compete for the same pool of funds. Under such circumstances, where social projects can inflict a loss to national consumption by diverting funds from more socially profitable private investments, public investments should yield at least the same return as private investments. It has been suggested that SOCC could be approximated by the marginal pre-tax rate of return on riskless private investments, for example the pre-tax rate on top-rated corporate bonds. Note that a relatively high SOCC would be expected in developing countries, as a result of scarcity of capital, compared to that in developed countries.

Attempts have been made to reconcile these two approaches; see Zhuang *et al.* (2007) for a survey of the theoretical and practical dimensions of the issue.

The full cost present value (*FCPV*) of a water supply project will usually be computed using prices that represent the national opportunity costs of the invested resources at a reference date, conventionally chosen to represent either the date of project completion or the beginning of the project life-cycle. This will require the use of a real annual discount rate, differing from a nominal annual discount rate by an annual rate of inflation. If historical cost data are used to estimate the cost components of the project, these past monetary values will have to be inflated to the overall price level of the reference date. This is done using a price index for the priced resource or, in the absence of such a specific price index, by means of a GDP deflator or a consumer price index.

Costs comparisons based on the *FCPV* indicator can also be performed using a cost indicator that is easier to interpret, namely the *full annual equivalent cost* (*FAEC*). The *FAEC* is defined as the constant annuity to be paid during the project life-cycle to refund the *FCPV* of the project at the annual real social discount rate *i*. The *FAEC* is computed by solving the following equation:

$$\sum_{t=1}^{T} \frac{FAEC}{(1+i)^{t-1}} = FCPV.$$

This solution leads to:

$$FAEC = FCPV \frac{i(1+i)^{T-1}}{(1+i)^{T} - 1}.$$

As explained in section 5.2, recurrent costs generally depend on the activity level at which the water supply system is operated. As a consequence, two variants of the *FCPV* and *FAEC* can be computed. The first variant assesses the recurrent costs by relying on the expected use of the design capacity of the project during its life-cycle, while the second variant assumes a 100% use of the design capacity throughout the project design lifetime. We call these two variants of a full cost indicator, *full cost at the expected use of design capacity*, and *full cost at the full use of design capacity*, respectively. For a project designed to provide an increasing annual level of services, the cost at the full use of design capacity overstates the cost of the expected use of design capacity by an amount determined by the time line for unused production capacity.

If the level of service provided by the available water supply technologies varies in time and across technologies, the *FCPV* or the *FAEC* are not the most suitable indicators to use for least-cost comparisons, because the value of these full cost indicators varies depending on the level of services provided. In such situations, a service or production indicator of the water supply system is needed to compute a cost measure per unit of service provided during a year.

In the general case where a water supply facility is not used at full capacity upon construction, but where its use increases gradually over time to meet the designed level of services, an appropriate definition of a unit cost is provided by the *average incremental cost* (*AIC*), based on the formula:

$$AIC = \frac{FCPV}{\sum_{t=1}^{T} \frac{S_t}{(1+i)^{t-1}}},$$

where S_t stands for the annual level of services provided in life-cycle year *t*. This formula provides a unit cost indicator calculated by dividing the *FCPV* of the water supply system by a measure of its life-cycle production which values the services provided in the future less than those produced at the present time, just as costs incurred in the future have a lower present value than those incurred at the present time. This way of measuring the life-cycle production of a water supply system that is operated over time at variable levels of capacity use, which we call the *life-cycle production present value* (*LCPPV*), expresses the present value of total production over the life-cycle of the project if the value of the services provided remains constant. Therefore, *AIC* may be viewed as the constant selling price of system production allowing the recovery of the total life-cycle costs of the water supply project.

Another unit cost indicator can be derived from the *FAEC* by dividing this annualized project cost by an annualized measure of its life-cycle production. Similar to the *FAEC* an *annual equivalent life-cycle production* (*AELCP*) can be defined as the constant annual level of production providing to system clients the same life-cycle production present value. Then, the *AELCP* is computed by solving the following equation:

$$\sum_{t=1}^{T} \frac{AELCP}{(1+i)^{t-1}} = LCPPV,$$

leading to:

$$AELCP = LCPPV \frac{i(1+i)^{T-1}}{(1+i)^{T}-1}.$$

Therefore, it appears that this intuitive unit cost indicator is identical to *AIC*:

$$\frac{FAEC}{AELCP} = \frac{FCPV}{LCPPV} = AIC.$$

To value the *opportunity cost of spare capacity*, we compute the *AIC* by assuming a full use of the design capacity throughout the project design lifetime, which we call the *average incremental cost at the full use of design capacity* (*AICF*). Then we compute the difference between the *average incremental cost at the expected use of design capacity* (*AICE*) and *AICF*, namely *AICE–AICF*. This value assesses an *opportunity cost of spare capacity* expressing the opportunity cost of spare capacity per unit of service provided by the project during its design lifetime. For a project designed to provide an increasing annual level of services, the *AICF* is expected to understate the *AICE* by an amount determined by the time path of the unused production capacity. Exceptions may occur when recurrent costs are prominent with respect to investment costs, and strongly related to the level of service provided or to the evolution over time of the price of resources used in recurrent activities. In such exceptional situations the *AICF* may overstate the *AICE*, implying a negative unit opportunity cost of spare capacity.

5.5.5 Designing service growth during the life-cycle of the water supply project

To quantify the services provided by a water supply system in each year *t* of its life-cycle, we consider three alternative indicators:

- the *size of the population served*, denoted by P_t;
- the *number of household water connections*, denoted by H_t;
- the *quantity of water supplied*, denoted by Q_t.

To design consistent life-cycle scenarios for these production indicators, we start by specifying independent scenarios for the population served and for the following two other variables:

- the *average size of the household served*, denoted by N_t;
- the *average per capita consumption of water of the population served*, denoted by q_t.

These last two variables allow us to derive (from the population-served scenario) consistent life-cycle scenarios for the number of household water connections and for the quantity of water supplied, simply by dividing P_t by N_t and by multiplying P_t by q_t, respectively. Thus we compute: $H_t = P_t/N_t$ and $Q_t = P_t q_t$.

The life-cycle scenarios of variables P_t, N_t and q_t can be entirely designed by the user, by setting the value of these quantitative indicators for each year of the project life-cycle. The scenarios may also be modelled in a more parsimonious way by means of the following formula:

$$X_t = X_1 + (X_{\theta+1} - X_1)F(t-1;\alpha,\beta,\theta),$$

where X_t denotes the variable to be modelled, X_1 its initial value when the water supply system starts to be used (beginning of year $t = 1$), $X_{\theta+1}$ its final value (beginning of year $t = \theta + 1$) corresponding to full capacity use of the system reached after $\theta \leq T$ full years of the project life-cycle (T full years), and $F(\tau;\alpha,\beta,\theta)$ a beta cumulative distribution function of the continuous time variable τ defined on the interval $[0;\theta]$.

This function expresses the shape of the time trend followed by variable X_t to reach, after θ full years, its final value $X_{\theta+1}$ from its initial value X_1. Therefore, the function depicts a *growth scenario* if $X_1 < X_{\theta+1}$ a *decline scenario* if $X_1 > X_{\theta+1}$ and a *steady state scenario* if $X_1 = X_{\theta+1}$.

The profile of this time trend is determined by the value of parameters α and β, which determine the shape of a beta cumulative distribution function and of its underlying density function, expressing the instantaneous rate of change (speed) of this time trend. By choosing appropriate values of these positive parameters α and β, a wide range of time trend profiles can be generated. A qualitative analysis of the shape of these profiles as a function of these parameters is presented in Annex V.

To summarize the detailed analysis of Annex V, five qualitatively different growth profiles of the cumulative distribution function $F(\tau;\alpha,\beta,\theta)$ can be generated, depending on the choice of one of the seven different time profiles that can be used to model the evolution of the instantaneous speed of growth during the water supply system's life-cycle of $\theta = T = 30$ years.

- As illustrated in Figure 5.5, by assuming a *bell-shaped* speed time profile ($\alpha > 1$, $\beta > 1$), namely a speed that first increases from 0 to a maximum value then decreases to zero, we generate an *S-shaped* growth profile, namely accelerated growth at the beginning of the life-cycle followed by a decelerated growth at the end of the life-cycle. These profiles are symmetrical about $\tau = \theta/2$ if $\alpha = \beta$.

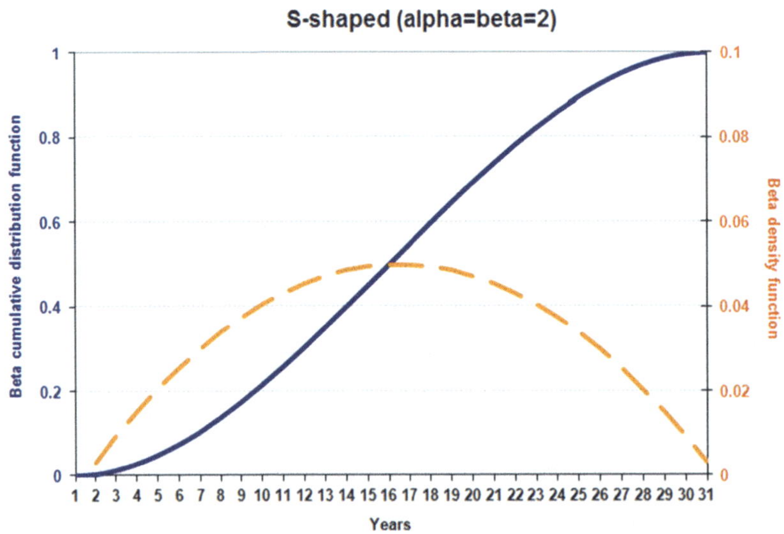

Figure 5.5 S-shaped growth profile.

- As illustrated in Figure 5.6, by assuming a *U-shaped* speed time profile ($0 < \alpha < 1$ and $0 < \beta < 1$), namely a speed that first drops from infinity to a minimum value then rises up to infinity, we generate a *rotated S-shaped* growth profile (about the segment joining the initial to the final growth trend value), namely decelerated growth at the beginning of the life-cycle followed by accelerated growth at the end of the life-cycle. These profiles are symmetrical about $\tau = \theta/2$ if $\alpha = \beta$.

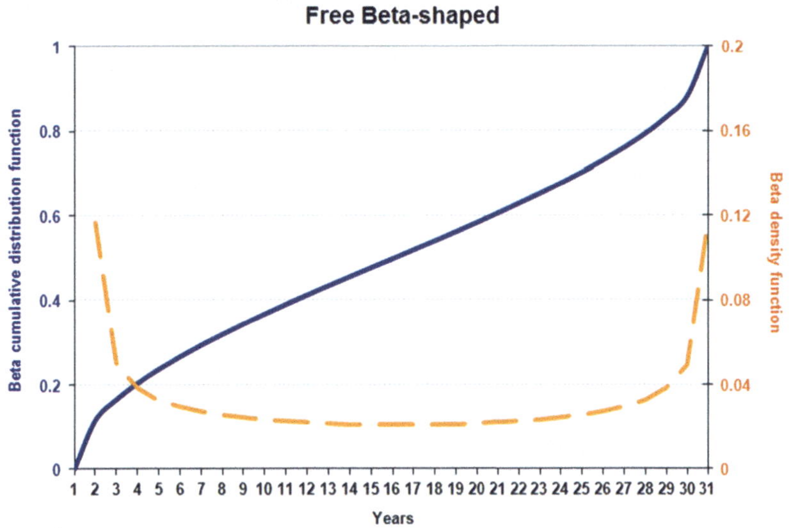

Figure 5.6 Rotated S-shaped growth profile.

- As illustrated in Figure 5.7, by assuming a *uniform-shaped* speed time profile ($\alpha = \beta = 1$), we generate a *linear-shaped* growth profile.

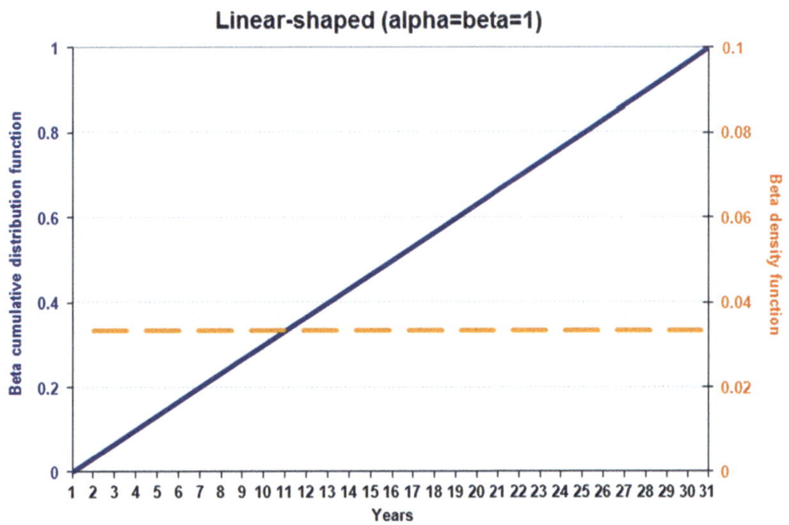

Figure 5.7 Linear-shaped growth profile.

- As illustrated in Figure 5.8, a *J-shaped* growth profile is generated by assuming an increasing speed time profile ($\alpha \geq 1$ and $0 < \beta \leq 1$), either *linear-shaped* ($\alpha = 2$ and $\beta = 1$), as in Figure 5.7, or *J-shaped* ($\alpha \geq 1$ and $0 < \beta < 1$ or $\alpha > 2$ and $\beta = 1$) with an increasing convex profile, or *rotated J-shaped* ($1 < \alpha < 2$ and $\beta = 1$) with an increasing concave profile, obtained by rotating the J-shaped profile about the segment joining the initial to the final profile value.

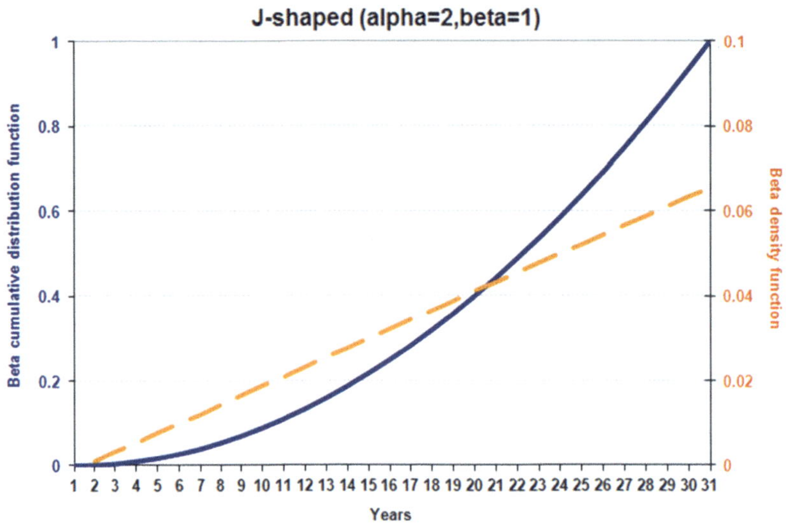

Figure 5.8 J-shaped growth profile.

- As illustrated in Figure 5.9, a *rotated J-shaped* growth profile is generated by assuming a decreasing speed time profile ($0 < \alpha \leq 1$ and $\beta \geq 1$), either *linear-shaped* ($\alpha = 1$ and $\beta = 2$) as in Figure 5.7, or *reverse J-shaped* ($0 < \alpha < 1$ and $\beta \geq 1$ or $\alpha = 1$ and $\beta > 2$) with a decreasingly convex profile, or *rotated reverse J-shaped* ($\alpha = 1$ and $1 < \beta < 2$) with a decreasing concave profile obtained by rotating the reverse J-shaped profile about the segment joining the initial to the final profile value.

The choice of one of these growth profiles to model a scenario of the services provided by a water supply system during its life-cycle depends on the kind of indicator used to quantify the system's production.

When the production indicator is represented by the services provided to the population served P_t measured in terms of *inhabitant-year* (services provided by the water supply system to an inhabitant during a full year), the growth profile of the cumulative distribution function $F(\tau;\alpha,\beta,\theta)$ should represent the growth path of the population served by the water supply system during its life-cycle. Demography shows that, in the very long run, the growth potential of a human population is bounded by the carrying capacity of the territory on which the population is settled. Therefore, in such a situation, the typical shape of a human population growth curve is that of an S-shaped profile. However, for shorter periods corresponding to the life-cycle of a water supply system, population growth corresponds to only a portion of such a long-run time path. In such a case, the use of another growth profile may be more appropriate. For example, a J-shaped profile will depict the initial accelerated growth of a population, a rotated J-shaped profile will depict the final decelerated growth of a population, and a linear-shaped profile will depict the population growth about the inflection point of an S-shaped profile, where the transition from the initial accelerated growth phase towards the final decelerated growth phase takes place.

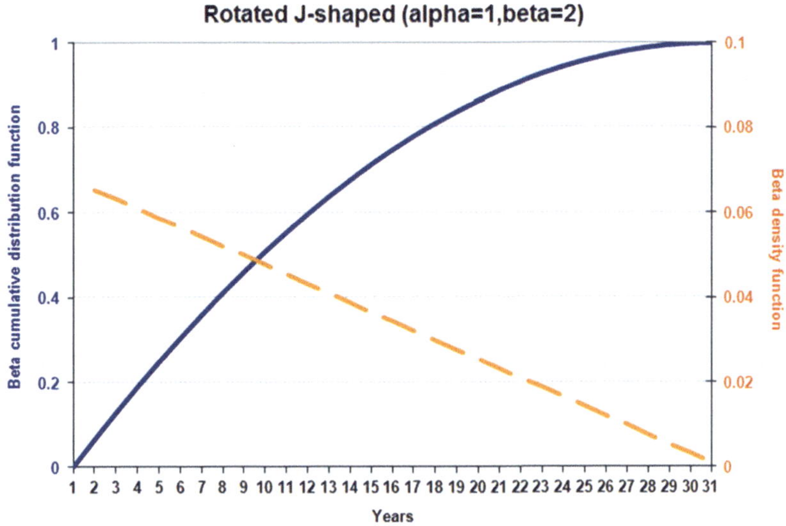

Figure 5.9 Rotated J-shaped growth profile.

When the production indicator is represented by the services provided to household water connections H_t measured in terms of *household-year* (services provided by the water supply system to a household during a full year), the previous population growth scenario should be supplemented by a scenario specifying the time path of the average size N_t of a household served by the system during its life-cycle, as indicator H_t is computed by dividing P_t by N_t. In the short run, this average size may be almost constant, but in the long run it will vary according to the demographic pattern of the households, by displaying either an increase, if the system is set up for a population of young households, or a decrease, if the system is set up for a population of mature households, for example in a rural settlement experiencing youth migration.

Similarly, if the production indicator is represented by the quantity of water supplied Q_t measured in terms of *litre/day-year* (litres of water per day supplied by the water supply system during a full year), the previous population growth scenario should be supplemented by a scenario specifying the time path of the average per capita daily consumption of water q_t of the population served by the system during its life-cycle, as indicator Q_t is computed by multiplying P_t by q_t. In the medium and long run, the per capita daily consumption of water may change as a consequence of the demographic composition of the population and especially of its socioeconomic development level. Therefore, before designing a life-cycle scenario for water consumption it is important to assess the effective demand of water by the community benefiting from the water supply project, as explained in section 4.3.

Chapter 6

Costing implementation

6.1 THE PROCESS OF COSTING A WATER SUPPLY PROJECT

Any attempt at costing a water supply project needs to take the following practical steps:

- outlining a project scenario;
- identifying the local constraints and risks faced by the project;
- selecting the appropriate improved water supply technologies compatible with the identified constraints and risks;
- identifying and collecting the relevant quantitative data needed to calibrate the level of water supply services to be provided and assess the life-cycle economic cost of the project for any identified locally appropriate water supply technology;
- computing consolidated cost indicators for any identified locally appropriate technology to perform least-cost analysis in order to identify the best technologies from an economic standpoint;

The operational design of a water supply project requires the definition of a baseline scenario complemented by a project scenario. Each of these scenarios has a cost attached to it for the provision of water supply. The project scenario reflects the various optimizations that can be done to better ensure reliable and sustainable access to safe drinking-water in sufficient quantities. Establishing the difference in cost between the two scenarios is referred to as 'incremental costing'.

Calculating the incremental cost of a project to improve drinking-water supply requires, in first instance, definition of the baseline scenario, which may depict:

- a community without a water supply infrastructure, but planning to build an infrastructure in the future;
- a community that wants to extend or improve its present water supply infrastructure.

The first scenario corresponds to a community that satisfies its basic needs for water by using natural, often unimproved sources (for example a spring, a pond or a river), but that plans to build a water supply system in the near future.

The second scenario corresponds to a community that currently uses a water supply infrastructure, but that wants to use another, better technology, for example by constructing a household connection system to replace or complement a system of public standpipes.

The distinction between these two kinds of project is important. In the first case, the incremental costing of the proposed water supply project requires a separate costing of both baseline and project scenarios – the difference is the incremental cost. In contrast, in the second case calculating the incremental cost of the water supply project can be performed by considering only the altered elements of the existing system and treating the value of the existing system as a 'sunk' cost. Therefore, depending on the project scenario selected, the carrying out the costing will require collecting all or only part of the information requested in the relevant questionnaires (Annexes II and IV).

The project scenario questionnaire (Annex II) seeks basic information on the qualitative and quantitative characteristics of the project, like: location, timeframe, demand scenario, type of technology.

A technical questionnaire and a socioeconomic questionnaire are also presented in Annex II. These two questionnaires aim to define the baseline scenario (the present situation of existing water supply technologies in the community) and to gather relevant data to help identify the constraints and risks faced by the water supply project (see section 4.2). The technical questionnaire is designed to collect basic data on the urbanization, hydrological and climatic characteristics of the community benefiting from the project, whereas the socioeconomic questionnaire seeks basic data on the demographic, economic and social characteristics of the community, as well as its potential contributions to the project. The assessment of other risks and constraints, such as financial, environmental, cultural and institutional aspects, may require collecting information not explicitly requested in these questionnaires.

All these data are essential inputs into the decision-making process, allowing for the identification and selection of the most appropriate water supply technologies that comply with the project objectives and address the risks and constraints faced by the project (see Figure 4.1).

To identify, collect and analyze cost data we have developed an Excel spreadsheet, which we call the *Water Supply Costing Processor* (WSCP). This WSCP enables the user-friendly collection and processing of the relevant quantitative information to compute the cost indicators presented in section 5.5 for any identified locally appropriate technology complying with the project scenario. These figures represent the basic information to perform a least-cost analysis allowing the identification of the most effective technologies from a cost standpoint.

The following section presents this tool, using the case study presented in Annex VI as an illustrative context. The case study concerns a project to supply potable water to a rural community in Peru.

6.2 THE WATER SUPPLY COSTING PROCESSOR (WSCP)

The WSCP runs with:

- Microsoft Excel 2003 or later (American version);
- Windows 95 or later;
- a personal computer with more than 4GB RAM memory.

The WSCP is composed of 31 spreadsheets, of which 21 are used to input data, 9 to provide output results and 1 contains a glossary of the technical terms displayed in the WSCP. The labels and the functions carried out by these WSCP spreadsheets are presented in tabular form in Figure 6.1 using the identifying colours of the spreadsheets.

6.3 AN INTRODUCTION TO THE USE OF WSCP

In this section we explain the use of the WSCP spreadsheets, taking the costing case study presented in Annex IV as an illustrative example. The case study concerns a project to supply potable water to a rural community in Peru.

Label	Function
Project design (1 sheet)	Selecting an improved water supply technology to be assessed and a time trend shape for the quantitative indicators used to design the life-cycle production growth of the water supply system.
Production scenarios (1 sheet)	Providing data to define the time trend shape of the quantitative indicators used to design the life-cycle production growth of the water supply system, when a free beta-shape profile or a free trend scenario is selected within the Project design menu of trend shapes, and displaying annual values and changes of these time trend indicators.
Population scenario (1 sheet)	Displaying a time trend graph of the size of the population supplied with water and a graph with the annual changes.
Household scenario (1 sheet)	Displaying a time trend graph of the number of household water connections and a graph with the annual changes.
Water scenario (1 sheet)	Displaying a time trend graph of the quantity of water supplied by the system and a graph with the annual changes.
Investment costs (7 sheets)	Identifying the main resources in local or imported materials, local or imported equipment, labour, incidentals and other resources required to set up the water supply technology, and providing data for costing these resources.
Maintenance costs (5 sheets)	Identifying the main resources in local or imported materials, local or imported equipment and labour required to maintain the water supply technology, and providing data for costing these resources.
Operation costs (5 sheets)	Identifying the main resources in local and imported materials, local or imported power services and labour required to operate the water supply technology, and providing data for costing these resources.
Other recurrent costs (1 sheet)	Identifying the main resources invested in administration, training, health and hygiene promotion, and education, and providing data for costing these resources.
Social costing (1 sheet)	Providing data for a social costing of the resources invested in the water supply project reflecting the national opportunity cost of these resources.
Costing summary (6 sheets)	Displaying the full and unit cost indicators of the resources invested in the whole water supply project and in each of its components.
Glossary (1 sheet)	Glossary of the technical terms displayed in the WSCP.

Figure 6.1 Labels and functions of the WSCP spreadsheets.

6.3.1 Selecting the technology and designing the use of the water supply project

Using the WSCP, the costing of a water supply project starts with the choice of an improved water supply technology and the design of its use over the project's life-cycle. This first task is performed using two WSCP spreadsheets.

The Project design spreadsheet requires the input of the following data.

- The choice of an **improved water supply technology** and its **design lifetime** (the expected duration of the water supply system, in years).

  A menu in the spreadsheet (displayed in Figure 6.2) lists the improved water supply technologies (see section 3.2 and Annex III for details).
- The **time trend shape** of each of the three quantitative indicators used to design the life-cycle production growth of the water supply system (see section 5.5), and the corresponding values of the three parameters used to quantify these trend profiles, namely: the **initial value** of the indicator (on the date the water supply system starts to be used), its **design value** (corresponding to the full capacity use of the water supply system) and the **trend duration** (number of full years to reach the design value).

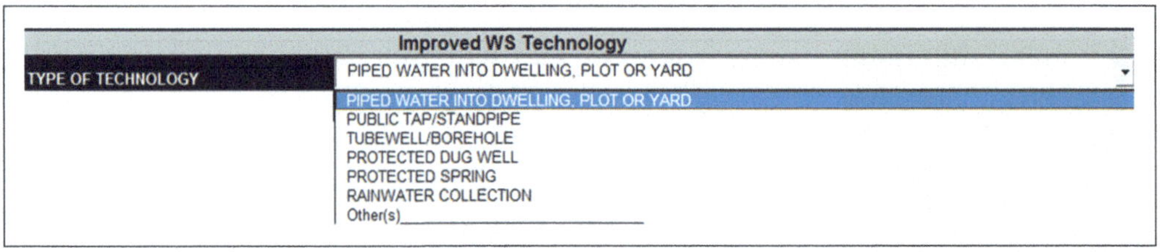

Figure 6.2 Menu of improved water supply technologies.

A menu in the spreadsheet (displayed in Figure 6.3) lists six pre-programmed profiles for the shape of the time trend, as described in section 5.5.

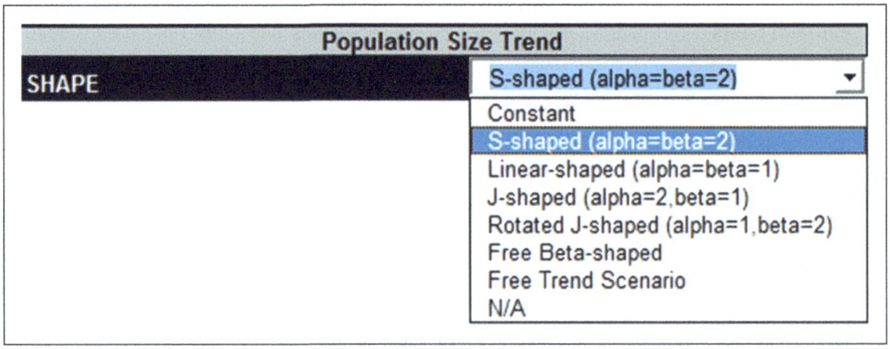

Figure 6.3 The menu of trend shapes.

- A UNIFORM profile, anticipating the indicator to be constant over the entire project life-cycle.
- A SYMMETRICAL S-SHAPED profile, anticipating an accelerated growth in the indicator value at the beginning of the life-cycle followed by a decelerated growth at the end of the life-cycle (defined by a beta cumulative distribution function with parameter values $\alpha = \beta = 2$).
- A LINEAR-SHAPED profile, anticipating a constant growth of the indicator value over the entire project life-cycle (defined by a beta cumulative distribution function with parameter values $\alpha = \beta = 1$).
- A J-SHAPED profile, anticipating an accelerated growth of the indicator value over the entire project life-cycle (defined by a beta cumulative distribution function with parameter values $\alpha = 2$ and $\beta = 1$).
- A ROTATED J-SHAPED profile, anticipating a decelerated growth of the indicator value over the entire project life-cycle (defined by a beta cumulative distribution function with parameter values $\alpha = 1$ and $\beta = 2$).
- A FREE BETA-SHAPED profile, defined by a beta cumulative distribution function which values of positive parameters α and β are freely set by the user.
  There is also a FREE TREND SCENARIO option. This can be entirely designed by the user, by setting the value of the quantitative indicator for each year of the project life-cycle.

The Production scenarios spreadsheet is used both to input the α and β parameters of the FREE BETA-SHAPED profile and the yearly values of the FREE TREND SCENARIO, as presented in Figure 6.4.

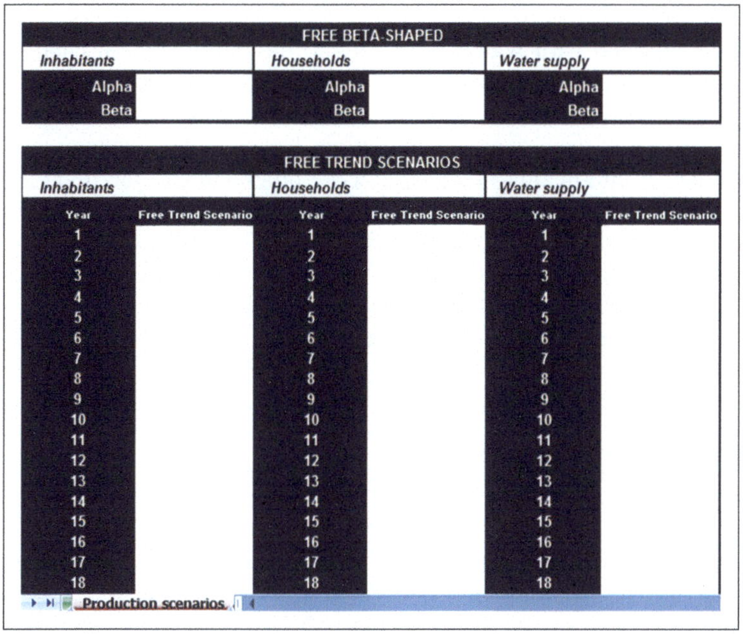

FREE BETA-SHAPED					
Inhabitants		Households		Water supply	
Alpha		Alpha		Alpha	
Beta		Beta		Beta	

FREE TREND SCENARIOS					
Inhabitants		Households		Water supply	
Year	Free Trend Scenario	Year	Free Trend Scenario	Year	Free Trend Scenario
1		1		1	
2		2		2	
3		3		3	
4		4		4	
5		5		5	
6		6		6	
7		7		7	
8		8		8	
9		9		9	
10		10		10	
11		11		11	
12		12		12	
13		13		13	
14		14		14	
15		15		15	
16		16		16	
17		17		17	
18		18		18	

Figure 6.4 Inputting a FREE BETA-SHAPED profile or FREE TREND SCENARIO in the Production scenarios spreadsheet.

Note also that for both the average household size and the per capita water consumption indicators, the trend shape menu includes a 'not available' (N/A) option. This prevents the WSCP from computing the number of household water connections or the volume of water delivered if no scenarios are available for these quantitative indicators of the production of the water supply system.

6.3.2 A practical example of applying the WSCP

To show how the WSCP is applied in practice, we use data from a water supply project in Guantánamo, Department of San Martín, Peru. This project was designed to provide drinking-water to a population of 50 families. Water is taken from a river located in a gorge, transported to a sedimentation pond, from where it flows through a slow sand filter to a reservoir. Water is piped from the reservoir into the distribution network and finally into households.

The initial population benefiting from this project was estimated at 300 inhabitants (6 persons per family) but the infrastructure was laid out to supply water to a design population of 408 inhabitants. The design population is expected to be reached after 7 full years of growth at an average annual rate of 4.5% (the historical annual growth rate over the decade 1993–2003). The expected lifetime of the system is estimated at 20 years.

We assume an average per capita consumption of water of 50 litres per day, throughout the lifetime of the project.

The Project design spreadsheet is therefore filled in as follows:

- Select from the menu of improved water supply technologies: PIPED WATER INTO DWELLING, PLOT OR YARD. Insert an expected **design lifetime** of 20 years.

- Select from the menu of time trend shapes a suitable option to design a scenario for the **size of the population supplied with water**. According to the information at hand to model the time trend of this production indicator, one can select a pre-programmed profile for the shape of this time trend or the FREE TREND SCENARIO option, allowing the user to set the value of the indicator for each year of the project design lifetime. For the water supply project in Guantánamo, only the quantitative information to calibrate a pre-programmed time trend profile is available. Thus, we select a suitable pre-programmed profile and calibrate it by inserting the values of the parameters: **time trend duration**, 7 years; **initial population**, 300 inhabitants; **design population**, 408 inhabitants. For the purpose of illustration we selected a SYMMETRICAL S-SHAPED profile (defined by a beta cumulative distribution function with parameter values $\alpha = \beta = 2$), assuming an accelerated growth of the population size at the beginning of the water supply system use followed by a decelerated growth at the end of the period of 7 years growth. Note that the FREE TREND SCENARIO could have been chosen on the assumption of population growth at an annual constant rate (the historical average annual rate of 4.5%). The use of this growth profile would require inserting in the Production scenarios spreadsheet the value of the size of the population for each year of the project lifetime.
- Select from the menu of time trend shapes, the N/A (not available) option for the **average household size** indicator, as no scenario for this variable was formulated. This will prevent the WSCP from computing a scenario for the **number of household water connections**.
- Select from the menu of time trend shapes: CONSTANT. This implies a steady trend in **average per capita water consumption**. Quantify this by calibrating the parameters: **time trend duration**, 20 years; **design value**, 50 litres per inhabitant per day.

The completed spreadsheet is displayed in Figure 6.5.

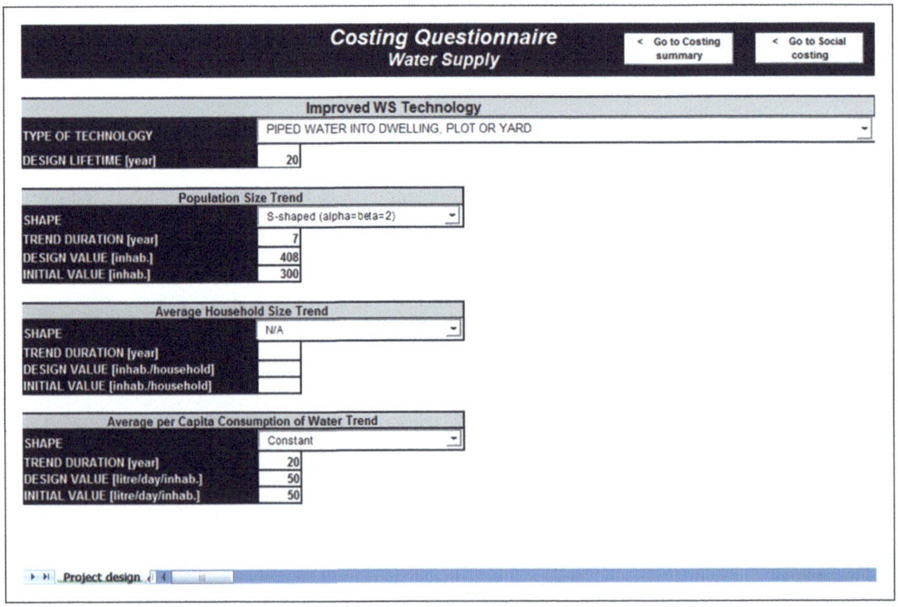

Figure 6.5 The Project design spreadsheet for the water supply to Guantánamo, Department of San Martín, Peru.

6.3.2.1 Displaying the project scenario

The scenarios of the indicators for the life-cycle production of the water supply project are displayed by the WSCP in two ways.

For each year of the project life-cycle, the Production scenarios spreadsheet shows figures of the following indicators.

- The **size of the population supplied with water**, along with the increment over the previous year.
- The **number of household water connections**, along with the increment over the previous year, if a scenario for the average size of the household served is included in the spreadsheet.
- The **quantity of water supplied**, along with the increment over the previous year, if a scenario for the average per capita consumption of water of the population served is included in the spreadsheet.

Figure 6.6 displays these figures for the population to be served and the quantity of water supplied by the water supply project in Peru.

Population served (P)			Water supplied (Q=P*q)		
Year	S-shaped (alpha=beta=2)	Speed S-shaped (alpha=beta=2)	q	Q	Speed Q
1	300	0	50	15000	0
2	306	6	50	15299	299
3	321	15	50	16071	771
4	343	21	50	17125	1055
5	365	23	50	18275	1149
6	387	21	50	19329	1055
7	402	15	50	20101	771
8	408	6	50	20400	299
9	408	0	50	20400	0
10	408	0	50	20400	0
11	408	0	50	20400	0
12	408	0	50	20400	0
13	408	0	50	20400	0
14	408	0	50	20400	0
15	408	0	50	20400	0
16	408	0	50	20400	0
17	408	0	50	20400	0
18	408	0	50	20400	0
19	408	0	50	20400	0
20	408	0	50	20400	0
21	408	0	50	20400	0

Production scenarios

Figure 6.6 Figures of the population served and the quantity of water supplied for the water supply project in Guantánamo, Department of San Martín, Peru.

A graphical presentation of the time trend scenarios of these production indicators is displayed in the following output spreadsheets.

- The Population scenario spreadsheet contains the time trend graph of the annual size of the population supplied with water, along with the time trend graph of the annual increment of this indicator over the previous year.
- The Household scenario spreadsheet contains the time trend graph of the annual number of household water connections, along with the time trend graph of the annual increment of this indicator over the previous year.
- The Water scenario spreadsheet contains the time trend graph of the annual quantity of litres of water per day supplied by the system, along with the time trend graph of the annual increment of this indicator over the previous year.

Figure 6.7 displays these graphs for the population to be served and the quantity of water supplied by the water supply project in Peru.

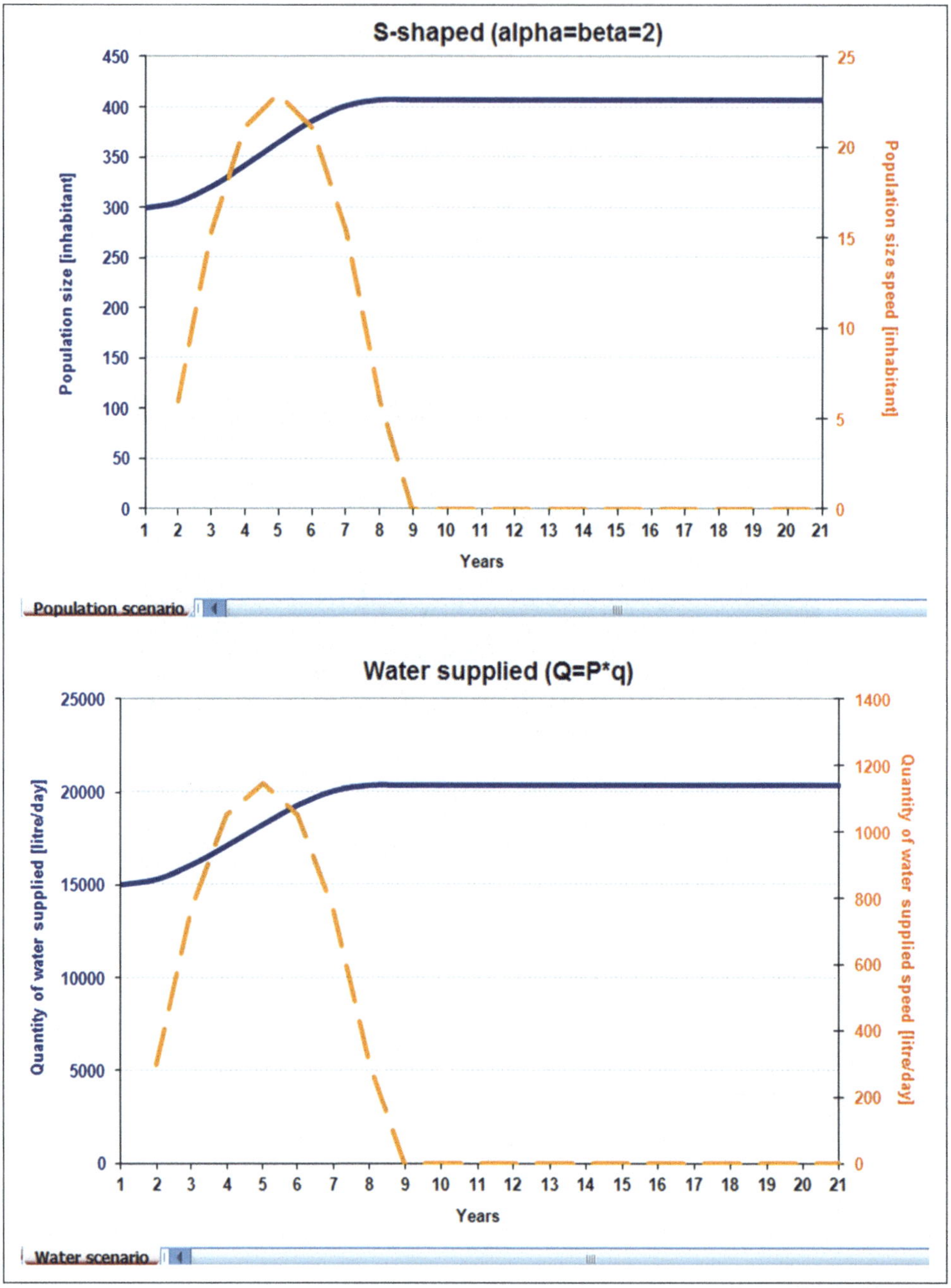

Figure 6.7 Graphs of the population served and the quantity of water supplied for the water supply project in Guantánamo, Department of San Martín, Peru.

6.3.2.2 Identifying and quantifying the resources invested in a water supply project

The second step in costing a water supply project consists in identifying the main resources required to install the water supply technology, and operate and maintain it during its design lifetime. To help identify the inputs necessary for implementing a particular water supply project, we have drafted a set of questionnaires encompassing all the activities potentially involved in implementing an improved water supply technology.

These costing questionnaires are discussed in section 5.4 and presented in Annex IV, based on the typology of costs discussed in section 5.2 and listed in Annex III. According to this typology, costs are first distinguished depending on whether they are incurred to set up the water supply infrastructure or to maintain and operate the infrastructure during its lifetime. This classification reflects the economic distinction between capital or investment inputs and current inputs. It generates the following basic cost typology, which is reflected in the costing questionnaires in Annex IV:

- investment costs
- maintenance costs
- operational costs
- other recurrent costs

These cost categories are then applied to each type of economic resource invested in the project, to generate the following 18 costing questionnaires:

- investment costs – local materials
- investment costs – imported materials
- investment costs – local equipments
- investment costs – imported equipments
- investment costs – labour
- other investment costs
- incidental investment costs
- maintenance costs – local materials
- maintenance costs – imported materials
- maintenance costs – local equipment
- maintenance costs – imported equipment
- maintenance costs – labour
- operational costs – local materials
- operational costs – imported materials
- operational costs – local power services
- operational costs – imported power services
- operational costs – labour
- other recurrent costs

Each of these costing questionnaires describes the use of a given resource according to three levels of aggregation. These levels are presented in the first three columns of the questionnaires, namely the **item**, **sub-item** and **input** levels, as described in section 5.4.

Once the resources involved in the realization of a water supply project have been identified, the next step is to look for data sources that provide the information needed to quantify the resources invested in the project. Depending on the available data sources, this quantification can be performed at a **disaggregated level** (input level breakdown) or at a more **aggregated level** (sub-item or item level breakdown). As discussed in section 5.4, the disaggregated level is recommended to provide clarity in the process of valuing economic resources in monetary terms and enable transferability of the data used to other settings, while the more aggregated levels have been devised for dealing with those situations where such detailed information is lacking.

6.3.2.2.1 The Investment costs spreadsheets

To complete the costing questionnaire for investment costs in **materials and equipment** at a disaggregated level, the following information is needed (to fill in columns 4 to 7) for each identified resource input:

- **acquisition cost** in local currency;
- **input quantity** and its unit of measurement (selected in a menu of pre-programmed units or provided by the user);
- **date of acquisition** in month and year (mm.yyyy).

When the aggregated level option is used, a physical measurement of the invested resources is not required. Obviously, these two levels of evaluation can be used simultaneously in costing a drinking-water supply project.

Whatever the level of aggregation adopted, the relevant costing information is put in by selecting the disaggregated level option displayed (along with the aggregated level option) on the upper-right corner of the investment cost sheet. By next selecting the aggregated level option, the input figures will be presented in a consolidated form at a sub-item or item level.

To illustrate this point, we display in Figure 6.8 the investment cost data used to quantify, at a disaggregated level, the resources in local material invested in the implementation of the tap and private connection infrastructures of the Guantánamo-San Martín water supply project. Figure 6.9 displays the same information but consolidated at the sub-item level by means of the aggregated level option.

Filling in the costing spreadsheet for investment costs in **labour** will require slightly different information:

- **hourly wage** in local currency;
- **input quantity** in hours;
- **date of wage payment** in month and year (mm.yyyy);
- **total wage cost** in local currency (when data are put in at an aggregated level).

Figure 6.10 displays the investment cost data used to quantify labour resources invested in the implementation of the tap and private connections infrastructures of the Guantánamo-San Martín water supply project. The available data sources did not allow this resource to be quantified at an input (disaggregated) level. Labour costs were therefore estimated at a more aggregated level by assessing, at sub-item level, the total wage cost.

The costing questionnaire sheet for **other investment costs** includes costs for preliminary studies, administration, promotion, and training and education/instruction of the project staff and users. These costs are usually assessed as a share of the construction costs, a quantity of worked hours or simply a lump sum. We present in Figure 6.11 the **assessment criteria** (selected in a pre-programmed menu) and data used to quantity these costs for the Guantánamo-San Martín water supply project.

The costing spreadsheet for **incidental investment costs** includes, in particular, labour costs for work supervision and engineering studies, as well as a lump sum for contingencies. These costs are not detailed at an input (disaggregated) level and therefore are displayed in the same way by the disaggregated and aggregated level options.

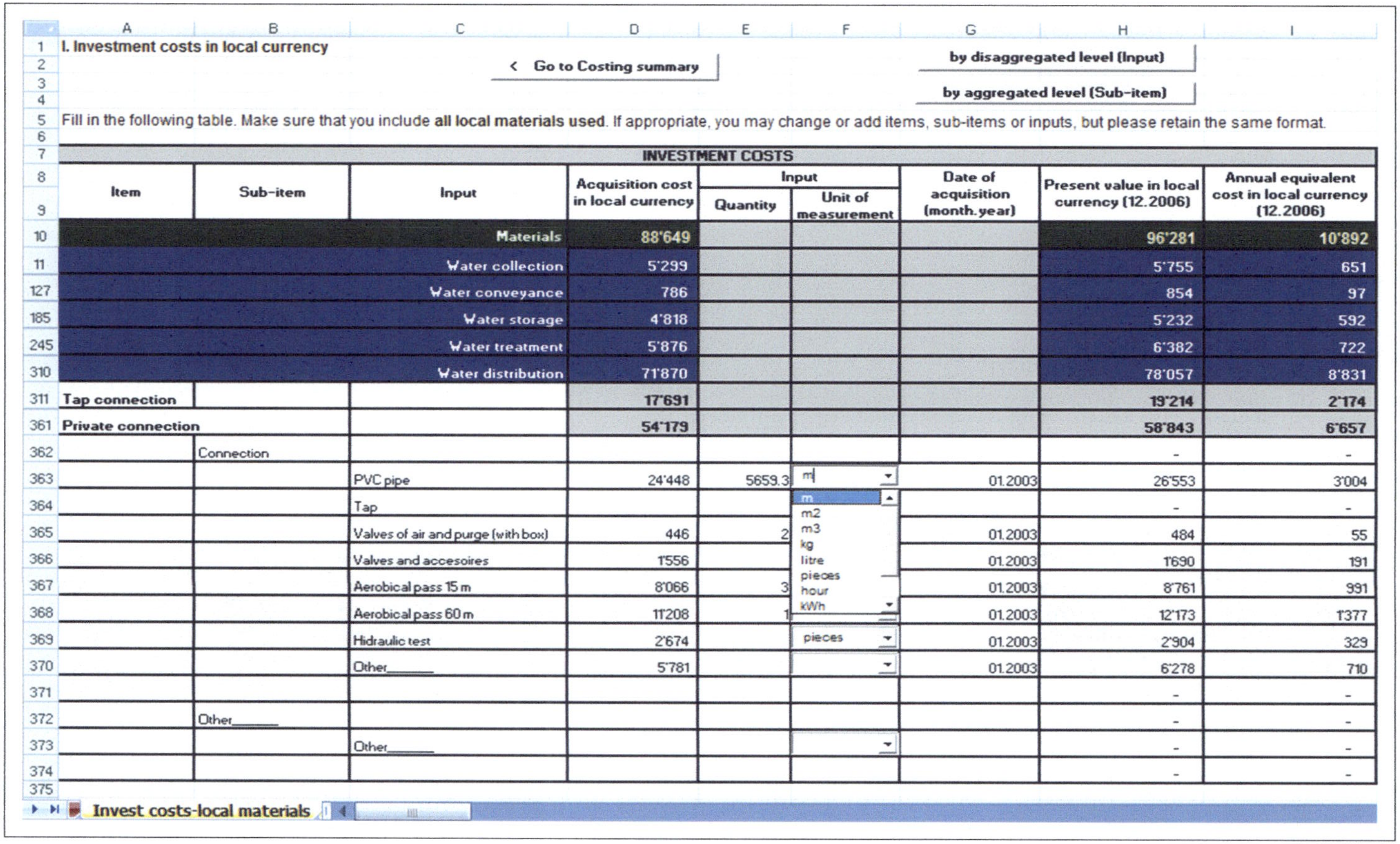

I. Investment costs in local currency

< Go to Costing summary | by disaggregated level (Input) | by aggregated level (Sub-item)

Fill in the following table. Make sure that you include **all local materials used**. If appropriate, you may change or add items, sub-items or inputs, but please retain the same format.

Row	Item	Sub-item	Input	Acquisition cost in local currency	Input: Quantity	Input: Unit of measurement	Date of acquisition (month.year)	Present value in local currency (12.2006)	Annual equivalent cost in local currency (12.2006)
10			Materials	88'649				96'281	10'892
11			Water collection	5'299				5'755	651
127			Water conveyance	786				854	97
185			Water storage	4'818				5'232	592
245			Water treatment	5'876				6'382	722
310			Water distribution	71'870				78'057	8'831
311	Tap connection			17'691				19'214	2'174
361	Private connection			54'179				58'843	6'657
362		Connection						-	-
363			PVC pipe	24'448	5659.3	m	01.2003	26'553	3'004
364			Tap			m		-	-
365			Valves of air and purge (with box)	446	2	m2	01.2003	484	55
366			Valves and accesoires	1'556		m3 / kg	01.2003	1'690	191
367			Aerobical pass 15 m	8'066	3	litre / pieces	01.2003	8'761	991
368			Aerobical pass 60 m	11'208	1	hour / kWh	01.2003	12'173	1'377
369			Hidraulic test	2'674		pieces	01.2003	2'904	329
370			Other_______	5'781			01.2003	6'278	710
371								-	-
372		Other_______						-	-
373			Other_______					-	-
374								-	-

Invest costs-local materials

Figure 6.8 The Investment costs spreadsheet for local materials for the Guantánamo-San Martín water supply project showing the input (disaggregated) level.

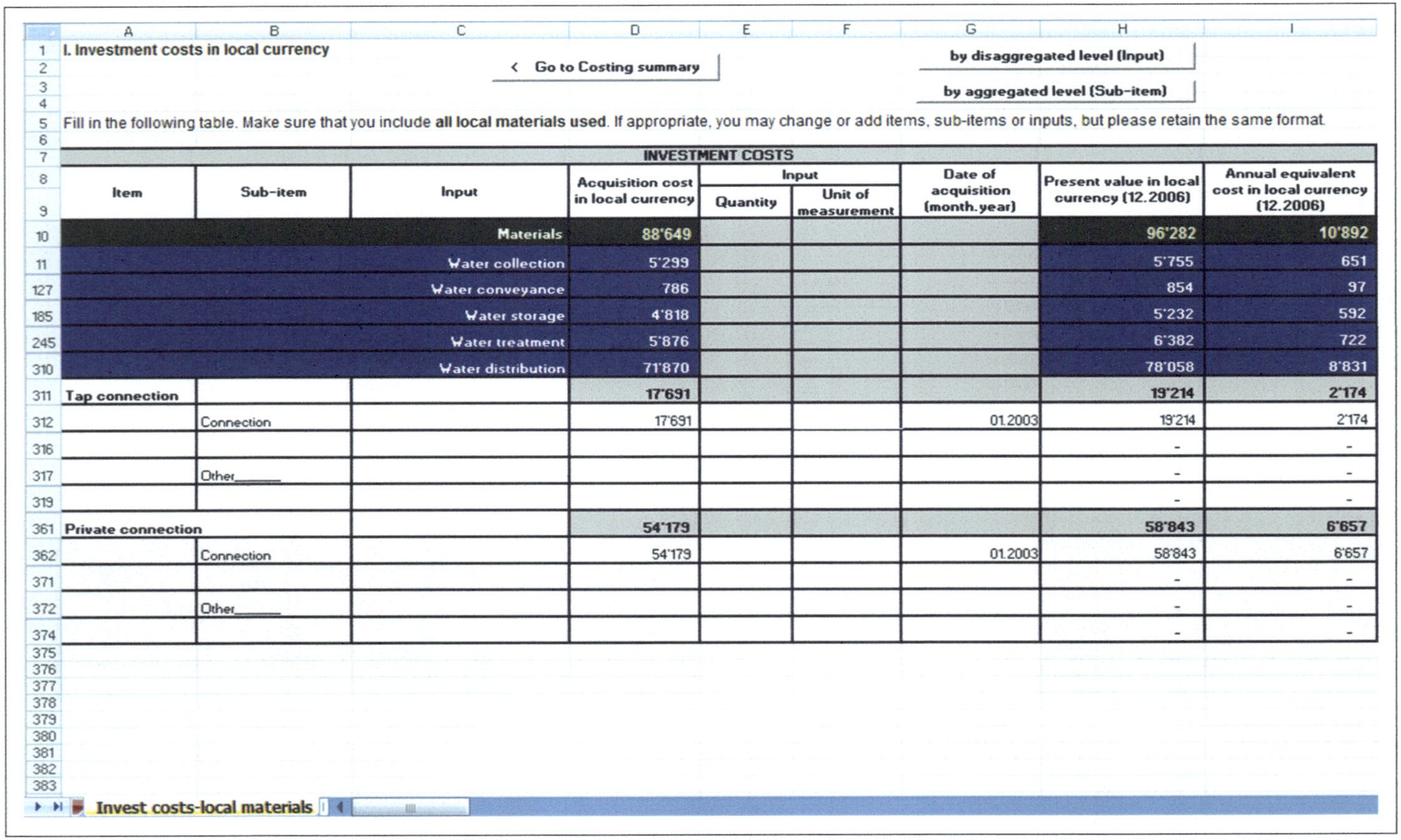

I. Investment costs in local currency

< Go to Costing summary | by disaggregated level (Input) | by aggregated level (Sub-item)

Fill in the following table. Make sure that you include **all local materials used**. If appropriate, you may change or add items, sub-items or inputs, but please retain the same format.

	INVESTMENT COSTS								
	Item	Sub-item	Input	Acquisition cost in local currency	Input: Quantity	Input: Unit of measurement	Date of acquisition (month.year)	Present value in local currency (12.2006)	Annual equivalent cost in local currency (12.2006)
10			Materials	88'649				96'282	10'892
11			Water collection	5'299				5'755	651
127			Water conveyance	786				854	97
185			Water storage	4'818				5'232	592
245			Water treatment	5'876				6'382	722
310			Water distribution	71'870				78'058	8'831
311	Tap connection			17'691				19'214	2'174
312		Connection		17'691			01.2003	19'214	2'174
316								-	-
317		Other_____						-	-
319								-	-
361	Private connection			54'179				58'843	6'657
362		Connection		54'179			01.2003	58'843	6'657
371								-	-
372		Other_____						-	-
374								-	-

Invest costs-local materials

Figure 6.9 The Investment costs spreadsheet for local materials for the Guantánamo-San Martín water supply project consolidated at the sub-item (aggregated) level.

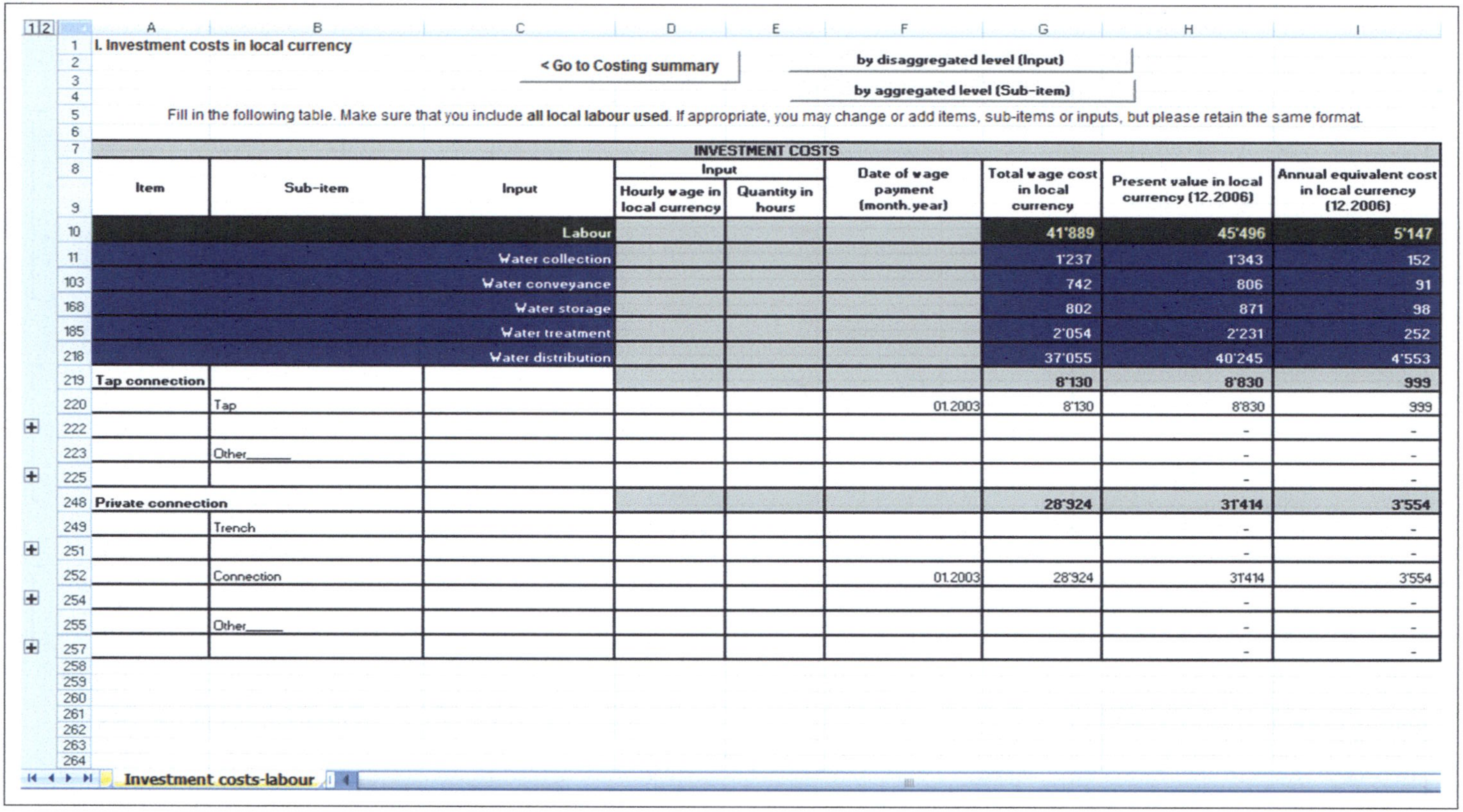

I. Investment costs in local currency

< Go to Costing summary | by disaggregated level (Input) | by aggregated level (Sub-item)

Fill in the following table. Make sure that you include **all local labour used**. If appropriate, you may change or add items, sub-items or inputs, but please retain the same format.

INVESTMENT COSTS								
Item	Sub-item	Input	Input: Hourly wage in local currency	Input: Quantity in hours	Date of wage payment (month.year)	Total wage cost in local currency	Present value in local currency (12.2006)	Annual equivalent cost in local currency (12.2006)
		Labour				41'889	45'496	5'147
		Water collection				1'237	1'343	152
		Water conveyance				742	806	91
		Water storage				802	871	98
		Water treatment				2'054	2'231	252
		Water distribution				37'055	40'245	4'553
Tap connection						8'130	8'830	999
	Tap				01.2003	8'130	8'830	999
							-	-
	Other_____						-	-
							-	-
Private connection						28'924	31'414	3'554
	Trench						-	-
							-	-
	Connection				01.2003	28'924	31'414	3'554
							-	-
	Other_____						-	-
							-	-

Investment costs-labour

Figure 6.10 The Investment costs spreadsheet for labour for the Guantánamo-San Martín water supply project showing the sub-item or item (aggregated) level.

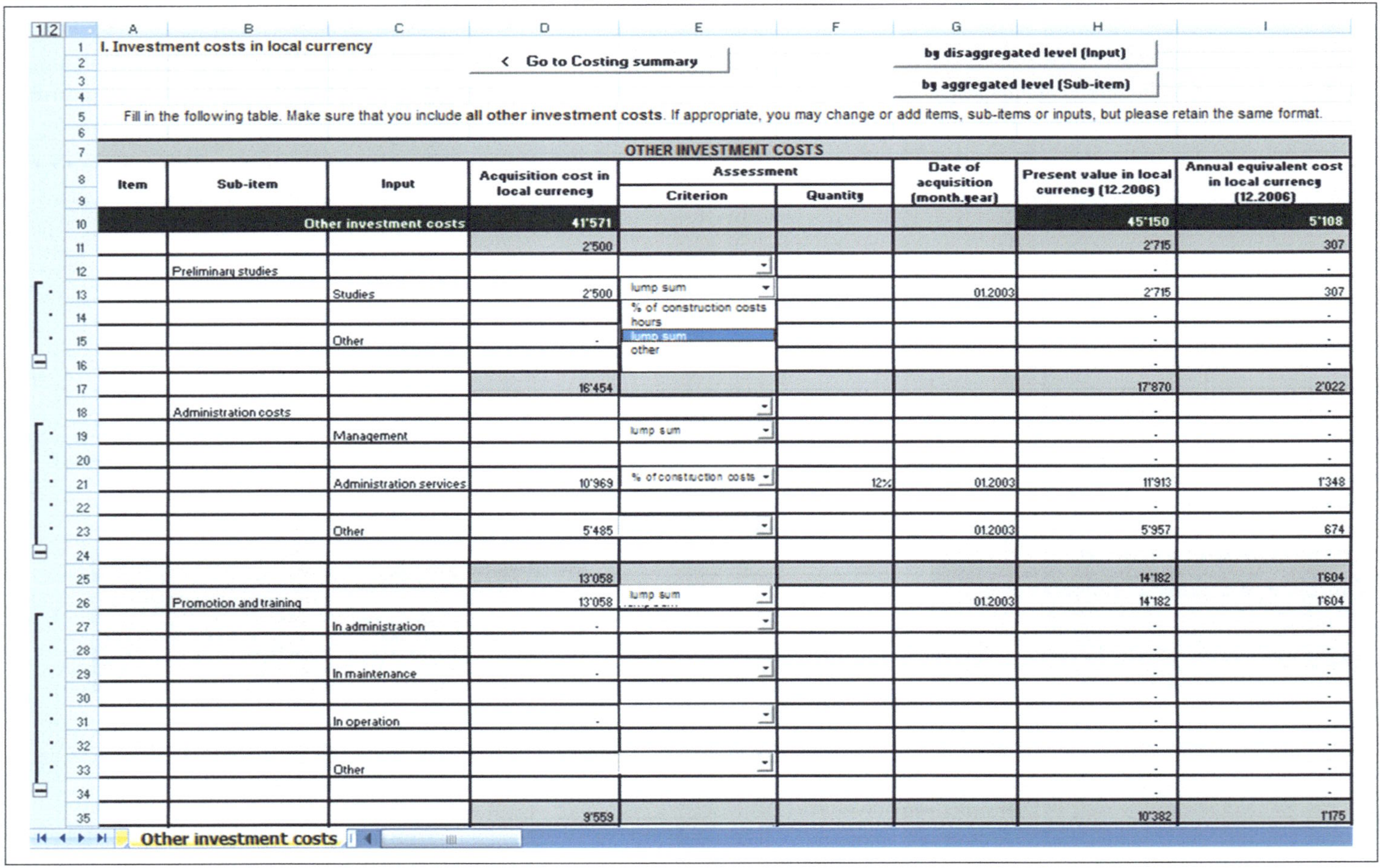

I. Investment costs in local currency

< Go to Costing summary

by disaggregated level (Input)

by aggregated level (Sub-item)

Fill in the following table. Make sure that you include **all other investment costs**. If appropriate, you may change or add items, sub-items or inputs, but please retain the same format.

	OTHER INVESTMENT COSTS								
	Item	Sub-item	Input	Acquisition cost in local currency	Assessment: Criterion	Assessment: Quantity	Date of acquisition (month.year)	Present value in local currency (12.2006)	Annual equivalent cost in local currency (12.2006)
10			Other investment costs	41'571				45'150	5'108
11				2'500				2'715	307
12		Preliminary studies						-	-
13			Studies	2'500	lump sum		01.2003	2'715	307
14					% of construction costs			-	-
15			Other	-	hours / lump sum			-	-
16					other			-	-
17				16'454				17'870	2'022
18		Administration costs						-	-
19			Management		lump sum			-	-
20								-	-
21			Administration services	10'969	% of construction costs	12%	01.2003	11'913	1'348
22								-	-
23			Other	5'485			01.2003	5'957	674
24								-	-
25				13'058				14'182	1'604
26		Promotion and training		13'058	lump sum		01.2003	14'182	1'604
27			In administration	-				-	-
28								-	-
29			In maintenance	-				-	-
30								-	-
31			In operation	-				-	-
32								-	-
33			Other					-	-
34								-	-
35				9'559				10'382	1'175

Other investment costs

Figure 6.11 The Other investment costs spreadsheet for the Guantánamo-San Martín water supply project showing the input (disaggregated) level.

6.3.2.2.2 The Maintenance costs and Operation costs spreadsheets

These costing spreadsheets are similar to those for investment costs, as far as the cost breakdown by system component, item, sub-item and input is concerned. The main differences lie in the nature of the input level, corresponding more often to a maintenance or operational intervention rather than to resource use, and in the timing of performing these recurrent activities.

The timing of these activities is specified by setting their **periodicity**, the period of time (in years) after which a recurrent activity must be repeated. For most of the operational activities the periodicity will be one year.

A recurrent cost may be **fixed** (independent of the level of services provided by the water supply facility) or **variable** (a function of the level of activity of the water supply facility). For simplicity, variable costs are estimated as a proportion of the cost of the recurrent activity assessed by assuming the water supply facility operates at its design capacity (**cost at design capacity**). The proportionality factor is determined by the level of services provided by the facility on the date the recurrent activity is performed. This activity level is measured by one of the production indicators selected by the WSCP user's in a pre-programmed menu. This menu includes, besides the fix cost option, the **size of the population supplied with water** (P), the **number of household water connections** (H) and the **quantity of water supplied** (Q). Obviously, the choice of one of these production indicators suppose that it has been previously defined by means of the Project design (for a modelled scenario using pre-programmed time trend profiles) or the Production scenarios spreadsheet (for a scenario entirely designed by the user).

The **real cost** (corrected for inflation) of a recurrent activity may evolve over time as a consequence of a change in the real price of some of the resources involved in carrying out the activity (for example energy prices). To deal with this issue, a trend scenario of the expected change in the real price of a resource or in the unit cost of an intervention can be calculated using the WSCP by specifying an **average annual rate of growth** of this variable for the expected lifetime of the project. If the real price of a resource or the real unit cost of an intervention is expected to remain the same over the system's life-cycle, this growth rate must be set equal to zero.

Filling in the costing spreadsheets for recurrent costs will therefore require the following information (to fill in columns 4 to 7) for each identified activity or resource input:

- **periodicity** in years round;
- **fixed or variable cost** (selected in a pre-programmed menu);
- **cost at design capacity** in local currency of the project reference date;
- **input quantity** and its unit of measurement (if available);
- **average annual growth rate of real unit cost**.

Figures 6.12 and 6.13 display the maintenance costs in local materials and labour assessed for the Guantánamo-San Martín water supply project at an aggregated level. Note that in small rural systems the separation of recurrent labour costs between maintenance and operation may be difficult as the same person is often in charge of both tasks.

Figure 6.14 displays the labour costs for operating the water treatment of the Guantánamo-San Martín water supply project assessed at an input (disaggregated) level.

6.3.2.2.3 The Other recurrent costs spreadsheet

This costing spreadsheet encompasses several recurrent costs that are necessary for the smooth functioning of the water supply system. The most important of these include the costs for administration, training in administration, maintenance and operation, as well as health and hygiene promotion and education. These recurrent costs are usually calculated as a share of the construction costs, a quantity of hours worked or simply a lump sum.

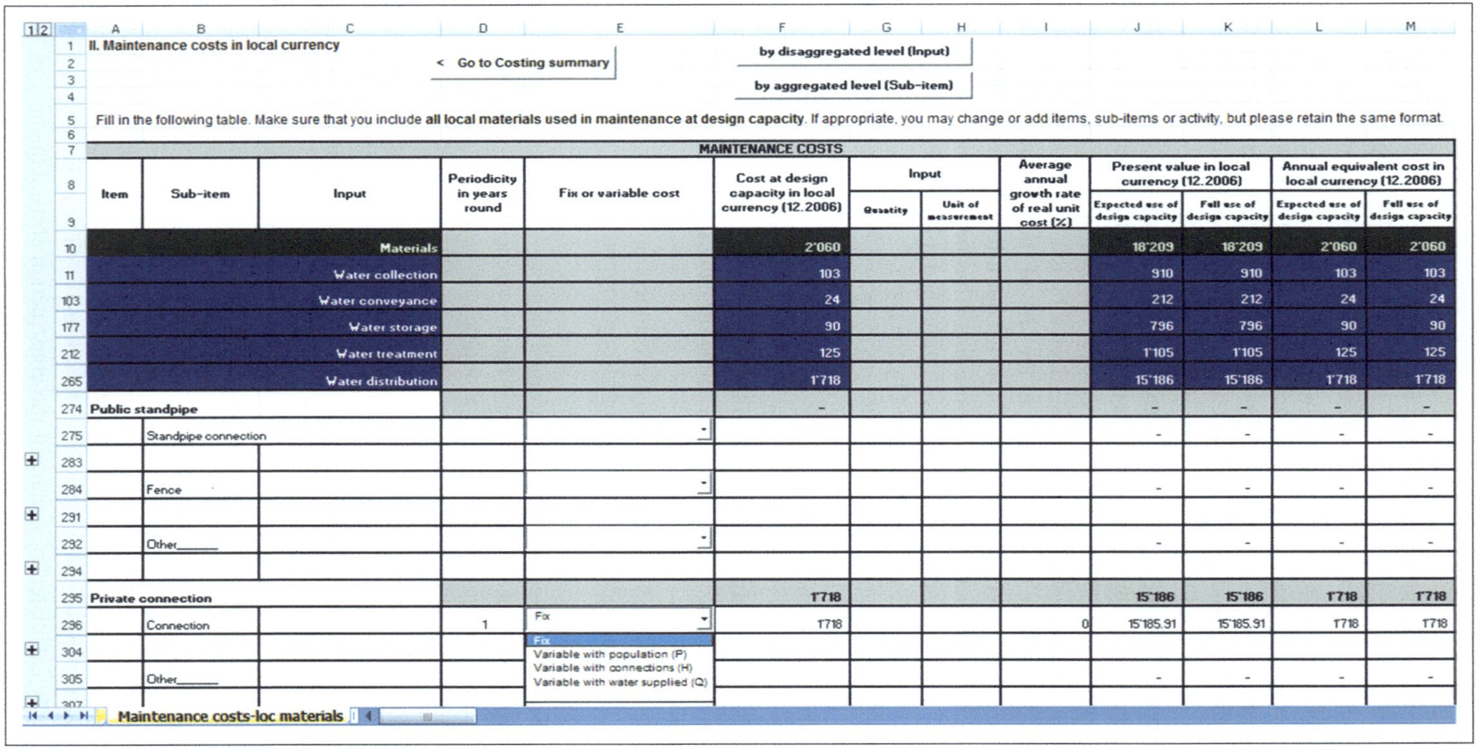

II. Maintenance costs in local currency

< Go to Costing summary | by disaggregated level (Input) | by aggregated level (Sub-item)

Fill in the following table. Make sure that you include **all local materials used in maintenance at design capacity**. If appropriate, you may change or add items, sub-items or activity, but please retain the same format.

MAINTENANCE COSTS

Row	Item	Sub-item	Input	Periodicity in years round	Fix or variable cost	Cost at design capacity in local currency (12.2006)	Input: Quantity	Input: Unit of measurement	Average annual growth rate of real unit cost (%)	Present value in local currency (12.2006): Expected use of design capacity	Present value: Full use of design capacity	Annual equivalent cost in local currency (12.2006): Expected use of design capacity	Annual equivalent: Full use of design capacity
10			Materials			2'060				18'209	18'209	2'060	2'060
11			Water collection			103				910	910	103	103
103			Water conveyance			24				212	212	24	24
177			Water storage			90				796	796	90	90
212			Water treatment			125				1'105	1'105	125	125
265			Water distribution			1'718				15'186	15'186	1'718	1'718
274	**Public standpipe**					-				-	-	-	-
275		Standpipe connection								-	-	-	-
283													
284		Fence								-	-	-	-
291													
292		Other_____								-	-	-	-
294													
295	**Private connection**					1'718				15'186	15'186	1'718	1'718
296		Connection		1	Fix	1'718			0	15'185.91	15'185.91	1'718	1'718
304					Fix / Variable with population (P) / Variable with connections (H) / Variable with water supplied (Q)								
305		Other_____								-	-	-	-

Maintenance costs-loc materials

Figure 6.12 The Maintenance costs spreadsheet for local materials for the Guantánamo-San Martín water supply project showing the sub-item or item (aggregated) level.

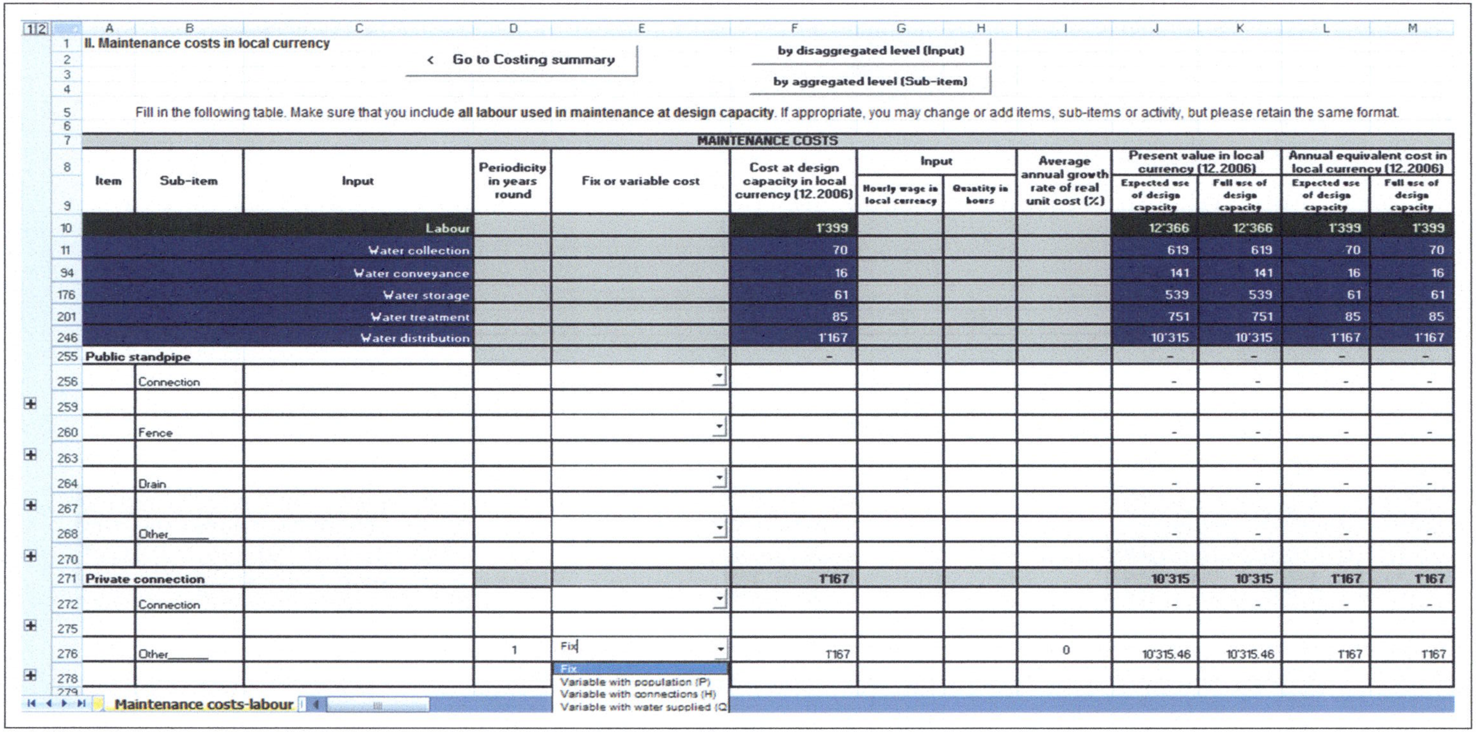

II. Maintenance costs in local currency

< Go to Costing summary | by disaggregated level (Input) | by aggregated level (Sub-item)

Fill in the following table. Make sure that you include **all labour used in maintenance at design capacity**. If appropriate, you may change or add items, sub-items or activity, but please retain the same format.

MAINTENANCE COSTS

Row	Item	Sub-item	Input	Periodicity in years round	Fix or variable cost	Cost at design capacity in local currency (12.2006)	Input: Hourly wage in local currency	Input: Quantity in hours	Average annual growth rate of real unit cost (%)	Present value in local currency (12.2006): Expected use of design capacity	Present value in local currency (12.2006): Full use of design capacity	Annual equivalent cost in local currency (12.2006): Expected use of design capacity	Annual equivalent cost in local currency (12.2006): Full use of design capacity
10			Labour			1'399				12'366	12'366	1'399	1'399
11			Water collection			70				619	619	70	70
94			Water conveyance			16				141	141	16	16
176			Water storage			61				539	539	61	61
201			Water treatment			85				751	751	85	85
246			Water distribution			1'167				10'315	10'315	1'167	1'167
255	Public standpipe					-				-	-	-	-
256		Connection								-	-	-	-
259													
260		Fence								-	-	-	-
263													
264		Drain								-	-	-	-
267													
268		Other_____								-	-	-	-
270													
271	Private connection					1'167				10'315	10'315	1'167	1'167
272		Connection								-	-	-	-
275													
276		Other_____		1	Fix	1'167			0	10'315.46	10'315.46	1'167	1'167
278													

Fix / Variable with population (P) / Variable with connections (H) / Variable with water supplied (Q

Maintenance costs-labour

Figure 6.13 The Maintenance costs spreadsheet for labour for the Guantánamo-San Martín water supply project showing the sub-item or item (aggregated) level.

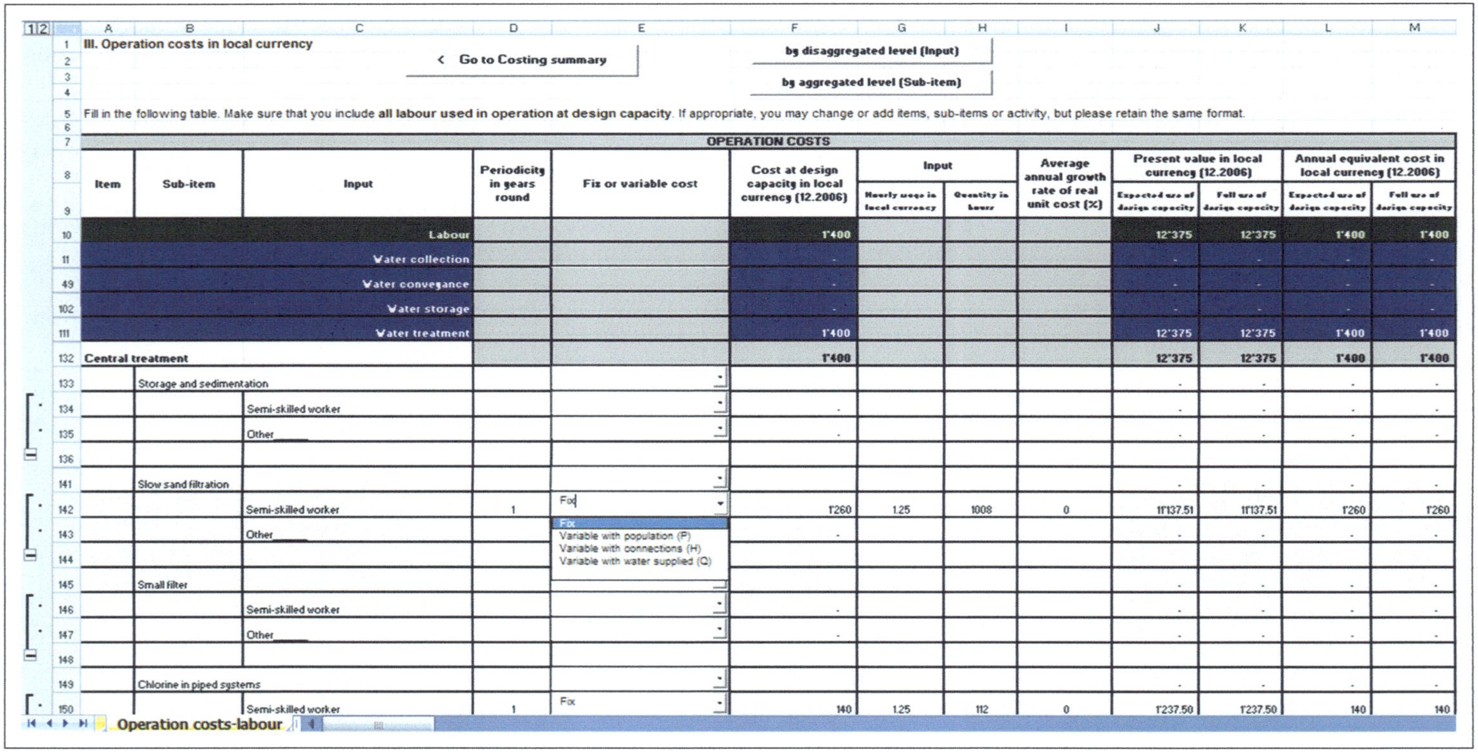

III. Operation costs in local currency

< Go to Costing summary

by disaggregated level (Input)

by aggregated level (Sub-item)

Fill in the following table. Make sure that you include **all labour used in operation at design capacity**. If appropriate, you may change or add items, sub-items or activity, but please retain the same format.

Row	Item	Sub-item	Input	Periodicity in years round	Fix or variable cost	Cost at design capacity in local currency (12.2006)	Input: Hourly wage in local currency	Input: Quantity in hours	Average annual growth rate of real unit cost (%)	Present value in local currency (12.2006): Expected use of design capacity	Present value in local currency (12.2006): Full use of design capacity	Annual equivalent cost in local currency (12.2006): Expected use of design capacity	Annual equivalent cost in local currency (12.2006): Full use of design capacity
10			Labour			1'400				12'375	12'375	1'400	1'400
11			Water collection			-				-	-	-	-
49			Water conveyance			-				-	-	-	-
102			Water storage			-				-	-	-	-
111			Water treatment			1'400				12'375	12'375	1'400	1'400
132	Central treatment					1'400				12'375	12'375	1'400	1'400
133		Storage and sedimentation								-	-	-	-
134			Semi-skilled worker			-				-	-	-	-
135			Other			-				-	-	-	-
136													
141		Slow sand filtration								-	-	-	-
142			Semi-skilled worker	1	Fix	1'260	1.25	1008	0	11'137.51	11'137.51	1'260	1'260
143			Other		Fix / Variable with population (P) / Variable with connections (H) / Variable with water supplied (Q)	-				-	-	-	-
144													
145		Small filter								-	-	-	-
146			Semi-skilled worker			-				-	-	-	-
147			Other			-				-	-	-	-
148													
149		Chlorine in piped systems								-	-	-	-
150			Semi-skilled worker	1	Fix	140	1.25	112	0	1'237.50	1'237.50	140	140

Operation costs-labour

Figure 6.14 The Operation costs spreadsheet for labour for the Guantánamo-San Martín water supply project showing the input (disaggregated) level.

Figure 6.15 displays the lump sum estimate of the administrative costs for the Guantánamo-San Martín water supply project.

6.3.2.3 Pricing the resources invested in a water supply project

The final step in costing a water supply project consists in valuing and discounting the quantified invested resources, using prices that represent the national opportunity costs of these resources (social costing).

To perform this task a Social costing spreadsheet has been designed to collect the relevant information for a social costing of the resources invested in the water supply project. This information is:

- the **local currency** used to value in monetary terms any cost component of the project;
- the **date of actualization** used as a reference date at which all the cost components of the project are to be valued and consolidated;
- the **real annual discount rate** used to compute the present value of each cost component of the project, as well as the project life-cycle production;
- the **shadow factors** for the **unskilled labour wage** and the **foreign rate of exchange** used to convert the actual market prices of unskilled labour, and imported materials and equipment into virtual competitive market prices, as explained in section 5.5;
- the series of past **consumer price index** values needed to transform historical cost data into cost values at the reference date.

As shown in Figure 6.16, the cost of the Guantánamo-San Martín water supply project was valued in the Peruvian currency, namely the Peruvian Nuevo Sol (PEN), at the shadow prices of 1 December 2006, by using a real annual discount rate of 11%, corresponding to the real social discount rate published by the Dirección General de Programación Multianual del Sector Público (2006) of Peru, and a shadow factor for unskilled labour of 0.49, reflecting the existence of a substantial amount of structural unemployment for unskilled labour in Peru. As no imported equipment or materials were used, no value was provided for the shadow factor for the rate of exchange.

The monthly series of the consumer price index was used to convert the investment cost components, input in the costing spreadsheets at the historical values of January 2003, to the PEN value at the reference date of 1 December 2006, conventionally chosen as the starting date of the use of the water supply system. These values are obtained by multiplying the historical values of January 2003 by a renormalized consumer price index having its base at the costing reference date of 1 December 2006. This renormalized index is displayed in the Social costing spreadsheet (see Figure 6.16).

All the recurrent costs must be valued at the value of the local currency at the reference date. For the Guantánamo-San Martín water supply project, this means calculating their value in PEN of 1 December 2006.

The last columns of the costing spreadsheets (see Figures 6.8 to 6.15) display, for each cost component, its **present value** and its **annual equivalent cost** at the reference date of the water supply project.

For the investment cost components, their present value simply corresponds to the valuation in local currency at the reference date, whereas for the recurrent cost components the present value is computed by consolidating the present values of the time sequence of expenditures incurred during the project design lifetime.

The **annual equivalent cost** of a cost component expresses the constant annuity to be paid during the project life-cycle to refund the present value of this component at an opportunity cost of capital taken as the real annual discount rate.

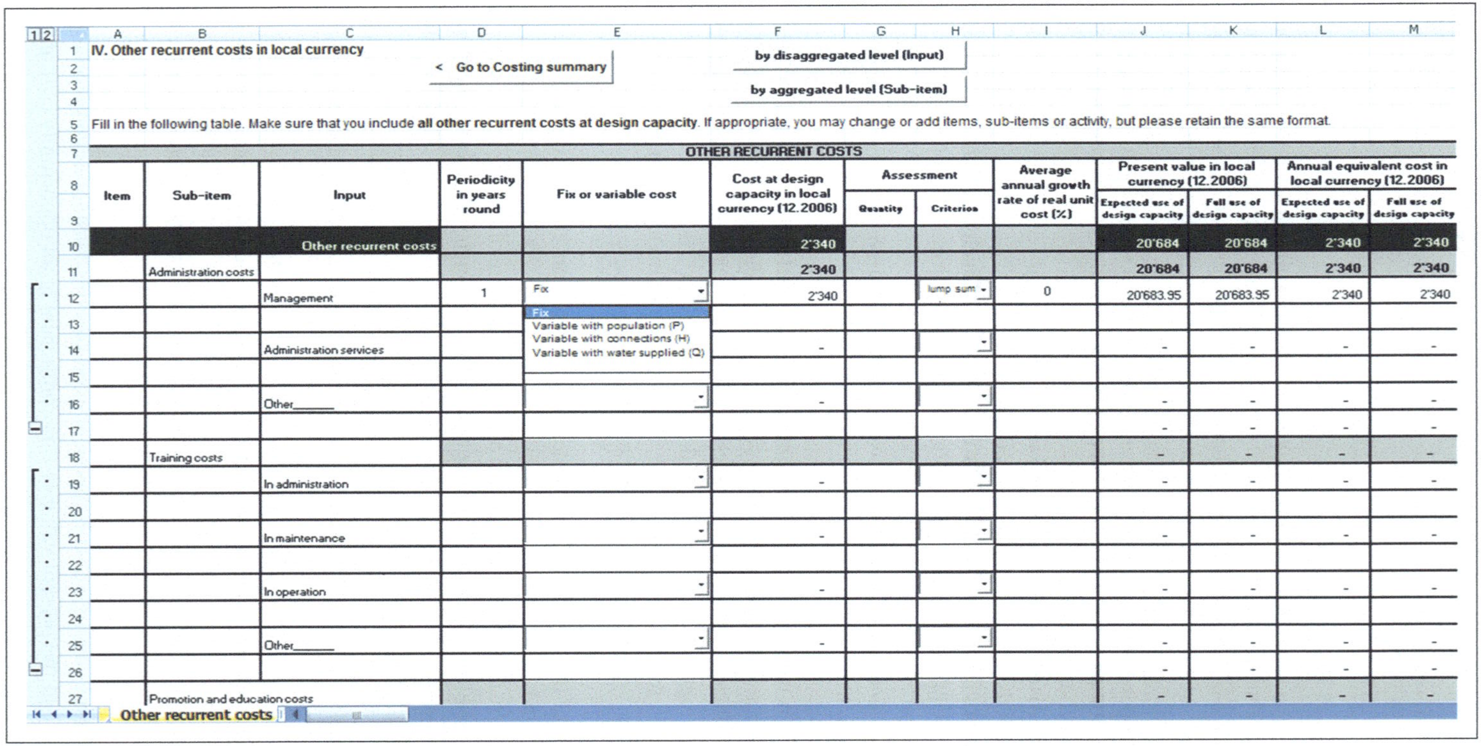

Figure 6.15 The Other recurrent costs spreadsheet for the Guantánamo-San Martín water supply project showing the input (disaggregated) level.

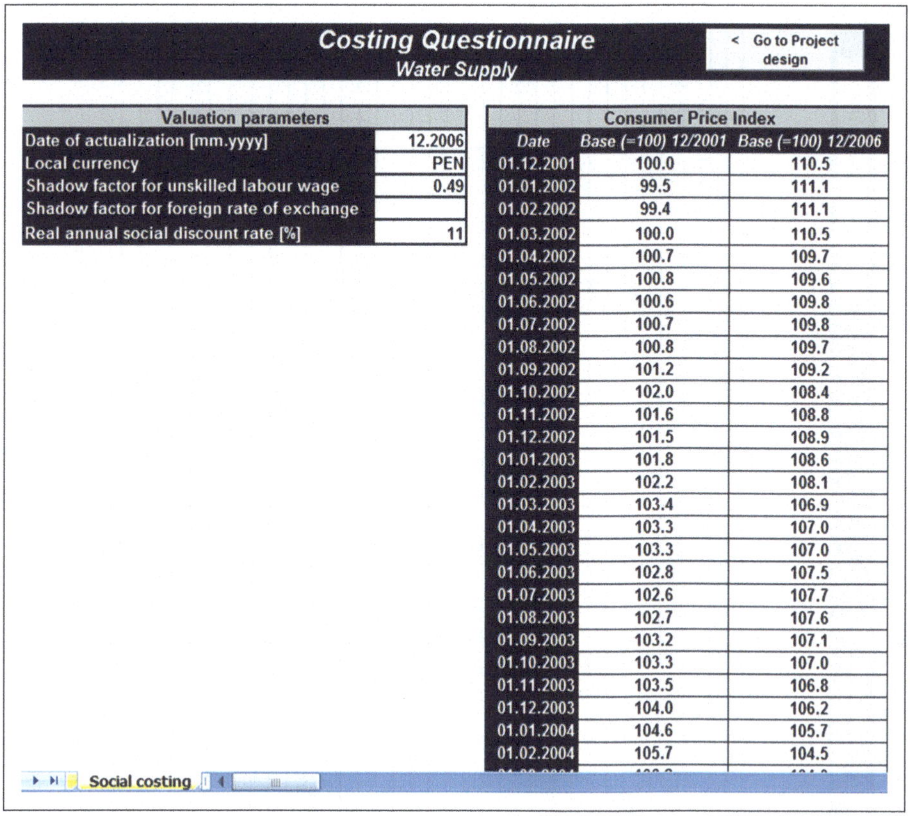

Costing Questionnaire
Water Supply

< Go to Project design

Valuation parameters	
Date of actualization [mm.yyyy]	12.2006
Local currency	PEN
Shadow factor for unskilled labour wage	0.49
Shadow factor for foreign rate of exchange	
Real annual social discount rate [%]	11

Consumer Price Index		
Date	Base (=100) 12/2001	Base (=100) 12/2006
01.12.2001	100.0	110.5
01.01.2002	99.5	111.1
01.02.2002	99.4	111.1
01.03.2002	100.0	110.5
01.04.2002	100.7	109.7
01.05.2002	100.8	109.6
01.06.2002	100.6	109.8
01.07.2002	100.7	109.8
01.08.2002	100.8	109.7
01.09.2002	101.2	109.2
01.10.2002	102.0	108.4
01.11.2002	101.6	108.8
01.12.2002	101.5	108.9
01.01.2003	101.8	108.6
01.02.2003	102.2	108.1
01.03.2003	103.4	106.9
01.04.2003	103.3	107.0
01.05.2003	103.3	107.0
01.06.2003	102.8	107.5
01.07.2003	102.6	107.7
01.08.2003	102.7	107.6
01.09.2003	103.2	107.1
01.10.2003	103.3	107.0
01.11.2003	103.5	106.8
01.12.2003	104.0	106.2
01.01.2004	104.6	105.7
01.02.2004	105.7	104.5

Social costing

Figure 6.16 The Social costing spreadsheet for the Guantánamo-San Martín water supply project.

In order to assess an **opportunity cost of spare capacity**, as defined in section 5.5, two variants of the cost present value and of the annual equivalent cost of each recurrent cost component are computed. The first variant, which we call a **cost at the expected use of design capacity**, assesses these costs by assuming the expected use of the design capacity of the project during its life-cycle, while the second variant, which we call a **cost at the full use of design capacity,** assumes a 100% use of the design capacity throughout the project design lifetime.

6.3.2.4 Displaying the costing of the water supply project

Once all the input data spreadsheets have been completed with the relevant information, the WSCP computes the following consolidated cost indicators for the water supply project, as described in section 5.5.

- The **full cost present value** (*FCPV*) at the reference date of the water supply project, broken down into its main components.
- The **full annual equivalent cost** (*FAEC*) of the water supply project, broken down into its main components.
- The **average incremental cost** (*AIC*), providing a unit cost calculated by dividing the full cost present value of the water supply system (or of its main components) by a measure of its life-cycle

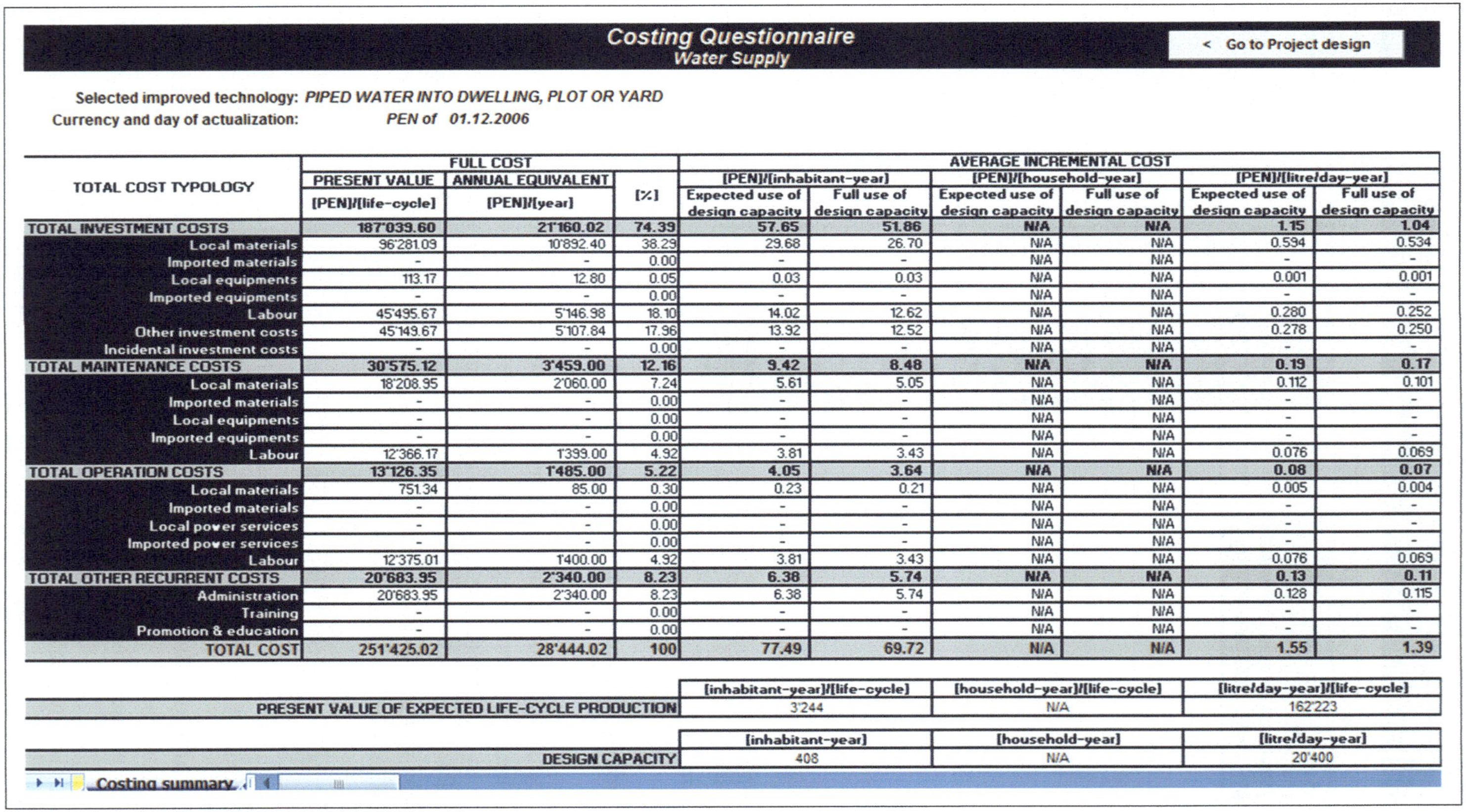

Costing Questionnaire
Water Supply

< Go to Project design

Selected improved technology: *PIPED WATER INTO DWELLING, PLOT OR YARD*
Currency and day of actualization: *PEN of 01.12.2006*

TOTAL COST TYPOLOGY	FULL COST			AVERAGE INCREMENTAL COST					
	PRESENT VALUE	ANNUAL EQUIVALENT	[%]	[PEN]/[inhabitant-year]		[PEN]/[household-year]		[PEN]/[litre/day-year]	
	[PEN]/[life-cycle]	[PEN]/[year]		Expected use of design capacity	Full use of design capacity	Expected use of design capacity	Full use of design capacity	Expected use of design capacity	Full use of design capacity
TOTAL INVESTMENT COSTS	**187'039.60**	**21'160.02**	**74.39**	**57.65**	**51.86**	**N/A**	**N/A**	**1.15**	**1.04**
Local materials	96'281.09	10'892.40	38.29	29.68	26.70	N/A	N/A	0.594	0.534
Imported materials	-	-	0.00	-	-	N/A	N/A	-	-
Local equipments	113.17	12.80	0.05	0.03	0.03	N/A	N/A	0.001	0.001
Imported equipments	-	-	0.00	-	-	N/A	N/A	-	-
Labour	45'495.67	5'146.98	18.10	14.02	12.62	N/A	N/A	0.280	0.252
Other investment costs	45'149.67	5'107.84	17.96	13.92	12.52	N/A	N/A	0.278	0.250
Incidental investment costs	-	-	0.00	-	-	N/A	N/A	-	-
TOTAL MAINTENANCE COSTS	**30'575.12**	**3'459.00**	**12.16**	**9.42**	**8.48**	**N/A**	**N/A**	**0.19**	**0.17**
Local materials	18'208.95	2'060.00	7.24	5.61	5.05	N/A	N/A	0.112	0.101
Imported materials	-	-	0.00	-	-	N/A	N/A	-	-
Local equipments	-	-	0.00	-	-	N/A	N/A	-	-
Imported equipments	-	-	0.00	-	-	N/A	N/A	-	-
Labour	12'366.17	1'399.00	4.92	3.81	3.43	N/A	N/A	0.076	0.069
TOTAL OPERATION COSTS	**13'126.35**	**1'485.00**	**5.22**	**4.05**	**3.64**	**N/A**	**N/A**	**0.08**	**0.07**
Local materials	751.34	85.00	0.30	0.23	0.21	N/A	N/A	0.005	0.004
Imported materials	-	-	0.00	-	-	N/A	N/A	-	-
Local power services	-	-	0.00	-	-	N/A	N/A	-	-
Imported power services	-	-	0.00	-	-	N/A	N/A	-	-
Labour	12'375.01	1'400.00	4.92	3.81	3.43	N/A	N/A	0.076	0.069
TOTAL OTHER RECURRENT COSTS	**20'683.95**	**2'340.00**	**8.23**	**6.38**	**5.74**	**N/A**	**N/A**	**0.13**	**0.11**
Administration	20'683.95	2'340.00	8.23	6.38	5.74	N/A	N/A	0.128	0.115
Training	-	-	0.00	-	-	N/A	N/A	-	-
Promotion & education	-	-	0.00	-	-	N/A	N/A	-	-
TOTAL COST	**251'425.02**	**28'444.02**	**100**	**77.49**	**69.72**	**N/A**	**N/A**	**1.55**	**1.39**

	[inhabitant-year]/[life-cycle]	[household-year]/[life-cycle]	[litre/day-year]/[life-cycle]
PRESENT VALUE OF EXPECTED LIFE-CYCLE PRODUCTION	3'244	N/A	162'223

	[inhabitant-year]	[household-year]	[litre/day-year]
DESIGN CAPACITY	408	N/A	20'400

Costing summary

Figure 6.17 The Costing summary spreadsheet for the Guantánamo-San Martín water supply project.

production. This cost indicator is calculated according to two variants of the use of the design capacity of the project, namely the **expected use**, and the **full use** throughout the project life-cycle, respectively. As explained in section 5.5, the difference between these two measures of the average incremental cost assesses an **opportunity cost of spare capacity** expressing the opportunity cost of spare capacity per unit of service provided by the project during its design lifetime.

These cost indicators are displayed in a series of Costing summary spreadsheets designed for costing the whole water supply project as well as each of its activities, namely:

- water collection
- water conveyance
- water storage
- water treatment
- water distribution

As illustrated in Figure 30, each costing summary presents the above cost indicators broken down by the cost typology described above in this section. The average incremental cost is calculated by using, as a quantitative measure of production, one or more of the three indicators used to design the life-cycle production of the water supply system, namely:

- the **life-cycle present value of the number of inhabitant–year** (inhabitants supplied with water during a full year);
- the **life-cycle present value of the number of household–year** (households supplied with water during a full year);
- the **life-cycle present value of the number of litre/day–year** (litres of water per day supplied during a full year).

As shown in Figure 6.17, the full cost present value of the Guantánamo-San Martín water supply project is evaluated (in PEN of 1 December 2006) as 251 425 [PEN]/[life-cycle] (Soles for the full life-cycle of the project), at the expected use of design capacity. Expressed as an annual equivalent cost, this full cost amounts to 28 444 [PEN]/[year]. Of these full costs, 75% arises from investment costs, 12% from maintenance costs, 5% from operation costs and 8% from administrative costs. This full cost present value is converted into an average incremental cost by dividing by the present value of its expected life-cycle production (valued at 1 December 2006), calculated as 3 244 [inhabitant–year]/[life-cycle] (inhabitants supplied with water during a full year for the full life-cycle of the project) or as 162 223 [litre/day–year]/[life-cycle] (litres of water per day supplied during a full year for the full life-cycle of the project). This leads to an average incremental cost (at the expected use of design capacity) of 77.54 [PEN]/[inhabitant–year] (PEN per inhabitant supplied with water during a full year) or 1.55 [PEN]/[litre/day–year] (PEN per litre of water per day supplied during a full year).

To assess the opportunity cost of spare capacity, the full cost present value and the present value of life-cycle production are recalculated by assuming a full use of the design capacity of the project. By dividing these two figures, an average incremental cost (at full use of design capacity) of 69.7 [PEN]/[inhabitant–year] or 1.39 [PEN]/[litre/day-year] is derived. Compared to the value of the average incremental cost computed at the expected use of design capacity, this indicates an opportunity cost of spare capacity of 7.8 [PEN]/[inhabitant-year] or 0.16 [PEN]/[litre/day-year].

Annex I

Drinking-water supply technologies

I.1 INTRODUCTION

This annex presents the wide range of drinking-water supply technologies. The objective is to describe existing technologies and highlight the factors that influence their costs, in order to help planners and decision-makers to select the most suitable option. The annex does not identify selection criteria, for which reference should be made to specialized engineering texts, such as Brikké and Bredero (2003) and SKAT (2000-2001).

Water supply technologies depend on local water sources and their adaptability to face up to different local constraints. A technology should, as far as possible, match people's needs, expectations, preferences and cultural habits. For a water supply technology to be acceptable to a community it should be affordable and convenient, and the technology should be easy to manage and maintain.

Relying mainly on Okun and Ernst (1987), and additionally on Brikké et al. (1997), SKAT (2000–2001) and WHO (1993), we first describe the different aspects of a drinking-water supply system, namely: the water sources and the activities of collection, conveyance, treatment, storage and distribution. Next, we present fact sheets summarizing the different technologies used to support the activities and comment on its operation and maintenance aspects, and its advantages and disadvantages. The fact sheets are based on work by Brikké et al. (1997), complemented by Okun and Ernst (1987), SKAT (2000–2001) and WHO (1993).

Each link in the chain of activities necessary to convey water from the water source to the final consumer involves a certain degree of technology. The details of these technical factors are not presented in this annex. In many countries excellent technical manuals are available, which may be useful in adapting the guidance given here to local features.

I.2 WATER SOURCES

The first step in designing a water supply system is to select a suitable source or combination of sources. The source must be capable of supplying enough water for the target community. The choice of a suitable source depends on local conditions. Here we deal with water sources commonly used in rural areas.

Often, a community uses several water sources for different purposes at different times of the year. Any planning for improvement should take into account the rationale behind existing patterns of water source use. Some sources may be more reliable or convenient or provide water that simply tastes better. If an 'improvement' means worse performance on any of these aspects, people may return to a traditional contaminated source.

The most important selection criterion for a water source is the quality of its water. Better quality of water will reduce the cost of treatment, thus making the source economically and technically more attractive. Other criteria for selecting a water source include yield, reliability, distance from the community that it is intended to serve, vulnerability to natural hazards such as flooding or freezing, and accessibility. Distance from the community may be less important if the water can be conveyed by gravity, but it is obviously a major factor if pumping is required. Besides the quantity and quality of the water, the final choice of the source will in practice depend on the cost of collecting the water (including the infrastructure and operation costs) and the funds available.

The possible types of water sources are:

- rainwater, direct precipitation catchments;
- groundwater from springs, infiltration galleries, shallow wells or deep wells;
- surface water from a stream, river, estuary or lake;
- water (raw or treated) purchased from a nearby city or community.

It is also possible to combine the use of two or more of the above sources.

I.2.1 Groundwater sources

Groundwater sources are the most commonly exploited for water supply because they are by far the safest and most practical. If available in adequate quantity and quality, groundwater is the source of preference because it requires less treatment than surface water, is easier to protect, and water supplies can be located closer to the targeted community. The disadvantages of groundwater are that it is often high in mineral content, it may be naturally contaminated with inorganic toxins such as arsenic or fluoride, and it usually requires pumping.

Different technologies are required for the different types of groundwater. Infiltration drains may be considered for shallow groundwater at medium depth. Tube wells are suitable for drawing water from deeper water-bearing ground strata. Dug wells are usually within local construction capabilities, whereas the drilling of tube wells requires more sophisticated equipment and considerable expertise.

Groundwater sources include:

- *Upland springs*
  Upland springs are almost ideal as they are least likely to be contaminated and they are often at a sufficient elevation to provide water by gravity feed. Because the recharge area is generally far from the population (usually more remote than other sources), they are more easily protected from man-made contamination.
- *Artesian springs and wells*
  An artesian spring discharges water under pressure. Artesian wells can be flowing or non-flowing. A flowing artesian well behaves like an artesian spring. In a non-flowing artesian well, the water level in the well is above the water table, but needs to be pumped. A major advantage of artesian sources is that they are not easily contaminated, because the aquifer is under pressure. The recharge area can be spoiled but, because it is generally far from the point of extraction, it tends to be cleansed in flowing underground. However, the yield of artesian sources and the likelihood that they are overdeveloped are major concerns. Excessive withdrawals may eliminate the pressure in the aquifer entirely.

- *Deep wells*
  A deep well can draw water from a free or a confined aquifer. Free aquifers are exposed to the surface, while confined aquifers have an impermeable overlying stratum. A confined aquifer is like an artesian aquifer without pressure. Wells in free aquifers are more subject to pollution than those in confined aquifers.
- *Infiltration galleries*
  Free groundwater flowing towards streams, estuaries or lakes from upland sources can be intercepted by infiltration galleries, which are perforated pipes laid perpendicular to the direction of flow. Infiltration galleries are useful in wetlands near coastal areas, where the deeper water is saline and the gallery can pick up the superficial fresh water. They are often used as river intakes to provide water of higher quality.
- *Shallow wells*
  Shallow wells are the most widely used sources of groundwater because their adequacy is most easily determined and they are cheaper to construct. Also, they can be fitted with hand pumps where powered pumps are not economically feasible. They are similar to infiltration galleries; one gallery is the equivalent of a line of shallow wells. However, shallow wells often suffer from deficiencies in both quantity and quality because their aquifers are shallow and close to the ground surface. Shallow wells are more likely to run dry in dry weather than other groundwater sources. They are also more prone to contamination of a microbial (inadequate sanitation) or chemical (for example percolation of agrochemicals) nature.

If groundwater is not available, or where the cost of digging a well is too high, it will be necessary to consider surface water as a source.

I.2.2 Surface water sources

Surface water originates mostly from run-off and rainfall water. It includes large rivers, ponds, lakes and small upland streams, which may originate from springs and collect run-off from watersheds. The quantity of run-off depends upon a large number of factors, the most important of which are the amount and intensity of rainfall, the climate, the vegetation, and the geological and topographical features of the area under consideration. The quality of surface water is governed by its content of living organisms, and by the amount of mineral and organic matter it has picked up in the course of its formation.

Surface sources should only be considered after possible groundwater sources have been explored. The advantage of surface sources is that they are generally more reliable in their yield, but current trends in climate change may undermine their reliability. Water from surface sources almost always requires treatment. The costs and difficulties associated with surface water treatment, particularly the day-to-day problems of operation and maintenance of water treatment plants need to be carefully considered before deciding to exploit surface water sources. However, surface water sources are preferred where groundwater is saline or high in arsenic, fluorides or other naturally occurring inorganic and potentially toxic contaminants.

Surface sources include:

- *Upland streams*
  Upland streams offer the best potential for surface supplies. Their watersheds are small, so they are least likely to be polluted and are most easily protected. Upland supplies also offer the potential for gravity supplies. However, dry weather flows may be inadequate to meet maximum water demands, so it may be necessary to create impoundments to provide for seasonal storage. Despite the isolation

of an upland source, a sanitary survey and water quality assessments in wet and dry seasons are necessary to provide a basis for watershed control activities and water treatment.

- *Lakes*
  Lakes can be excellent sources of water. If located at high elevation they can provide gravity flow. Their main disadvantage is the potential for impaired quality as a result of activities on the watershed and on the lake itself. A sanitary survey and water quality analyses are critical. Polluted lakes may present more problems than rivers because rivers tend to cleanse themselves when the pollution abates, while lakes may require a long time to overcome the effects of polluting discharges.
- *Rivers*
  Rivers offer convenient sources of supply for small communities, because they are almost always large enough to meet quantitative needs. However, rivers are generally the least appropriate sources because they are often of poor quality and small communities are not in a position to exert much control over upstream pollution. Also, intakes on large rivers are likely to be costly. Pumping costs are likely to be high if rivers tend to be at a relatively low elevation. Where river water is the only option, it may be appropriate to consider taking water from a larger community nearby in order to offset the cost of water treatment.

I.2.3 Regional supply

Water supply projects exhibit significant economies of scale. If a water source is adequate, developing it for twice or three times the capacity reduces the unit cost substantially. Hence, the development of a source for two, three or more communities provides significant economies. Such joint development of a common source may make it economically feasible to develop a higher quality source, for example a source at a greater distance that would be too costly for an individual community to develop.

Depending upon local circumstances, such as the geographical layout of the communities and the topography, it may be feasible to build a common treatment facility and to operate the water supply systems for several communities as a single entity. Such regionalization will produce additional economies as well as increasing operating effectiveness through the employment of better-qualified staff.

Another option is to take water from a nearby community that has developed a source and facilities larger than it needs. Such an approach may be economically attractive to both the communities. One community can save on construction costs, while the other community can reap income from its investment in resources that are otherwise not being used.

I.2.4 Combined use

Where a single source is not expected to be adequate year round, consideration will need to be given to using two sources combined. The most promising approach is the combined use of surface and groundwater. In many parts of the world, streams tend to dry up during the dry season, although they may provide an adequate water supply during the rest of the year. If it is not feasible to build an impounding reservoir to provide seasonal storage, a groundwater source may be used during the dry period. During wet weather, water is drawn from the stream and the groundwater aquifer is allowed to recharge. In this case, the aquifer provides seasonal storage.

I.3 WATER COLLECTION TECHNOLOGIES

The activities to be undertaken for water collection are determined by the expected needs of the users and the potential yield of the source. There are many ways to collect water, such as catchment systems, wells and intakes. The idea is to collect or extract water and store it before treatment.

The collection activity depends on the permeability and the production capacity of the water source. The importance of these factors varies considerably depending on local conditions. A hydrologist may need to be consulted to evaluate these factors if experience with wells in the vicinity does not provide enough guidance.

I.3.1 Catchment systems

Catchment systems are a way of collecting water for storage. The most common examples are rainwater catchments, storage dams and spring water catchments.

Rainwater catchments collect rain on the roof of a house using gutters and down-pipes, leading it to one or more storage containers, ranging from simple pots to large Ferro-cement tanks.

In storage dams, water from surface sources can be gathered for human use. Water is caught and stored behind the dam or diverted to a separate reservoir. Important factors in the planning of dams are annual rainfall and evaporation patterns, present use, and run-off coefficients of catchment area and building site. The user community can take the water directly from the dam, or the water can be treated and distributed in a network system.

Spring water catchment systems catch and protect ground water flows at the points where these arrive at the surface, to facilitate extraction. Two main categories of spring water catchment exist: in *gravity springs,* water flows on a natural underground slope to the surface, where it flows horizontally out onto the ground; in *artesian springs,* water is trapped between impervious layers and is forced vertically to the surface under pressure.

I.3.2 Wells

Wells are generally classified according to their method of construction. For water supplies wells may be dug by hand, driven, jetted, bored or drilled. Selection of the type of well depends upon the depth of the aquifer, the soil type, and the construction equipment and financial resources available. All wells have four basic parts: casing, intake area, wellhead and water lifting device.

- *Hand-dug wells*

  Hand-dug wells are the most common method of extracting water from the ground. The construction requires neither special equipment nor special skills. They can be risky to build, however, unless the necessary knowhow is available. Hand-dug wells can be constructed cheaply, with local equipment and materials. Their most important advantage is that water can be drawn from such wells using a bucket and rope, if the cost of a purchasing or maintaining a pump is out of reach. The depth of the wells is rarely more than 10 metres and they are generally 1 to 1.3 metres in diameter. The yield of dug wells depends mainly on the permeability of the aquifer, the well depth and the soil conditions. Enlargement of the diameter increases the yield slightly. For example, doubling the diameter augments the yield by less than 10%, but it does provide storage.

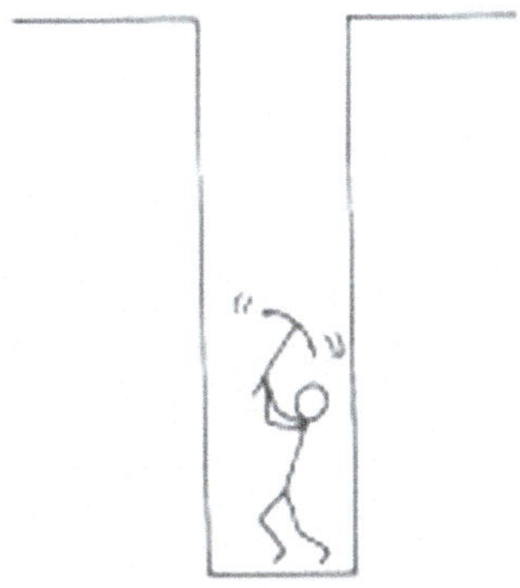

Source: SKAT (2000–2001)

- *Driven wells*
  To build a driven well, a pointed metal tube (known as a well point) is fitted to the end of a pipe and driven into the ground, usually by hitting it with a heavy hammer, until the point reaches the water-bearing level. Driven tube wells are a suitable option in soft ground, to depths of up to 15 m, but are inappropriate in hard rock formations and areas where many boulders or other hard obstacles are encountered in the ground. In clay soil, the screen opening may become clogged with clay during the driving. The feasible diameter ranges from 30 to 50 mm. The yield of driven wells is normally from 5 to 100 m^3 per day, but may be more in very permeable soils.

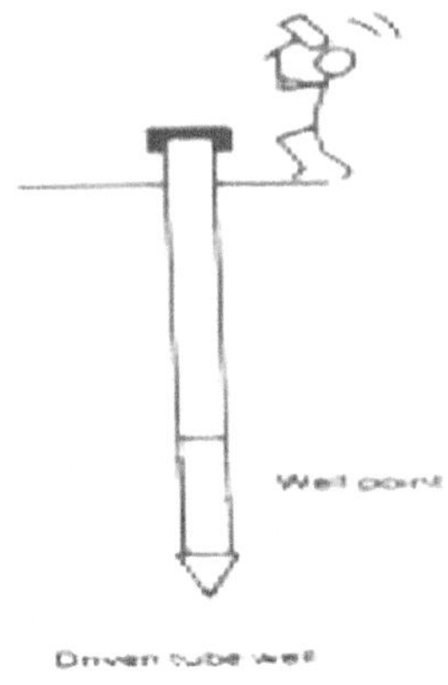

Source: SKAT (2000–2001)

- *Jetted wells*
  In jetted wells, as the name implies, a jet of water is used to help to develop the borehole. This requires specialised equipment and a plentiful supply of water, which limits the application. A pump with a capacity of approximately 500 litres per minute at a pressure of 3 to 5 atmospheres is necessary to create the jetting stream. The application of jetting stream technology is limited to unconsolidated soil formations without large boulders. Jetting of wells is possible to great depths, up to 50 m or more. The diameter of the jetting pipe generally ranges from 30 to 50 mm, but larger jetting pipes may be used. The simplest method of bringing up the water, known as the 'palm and slugger', consists of a pipe that is moved up and down by a lever, with a person's hand on the top of the pipe. The hand is used as a valve to pump water upwards, closing as the pipes rises and opening as it falls. The yield of jetted wells has the same range as that of driven wells: 5–100 m^3 per day.

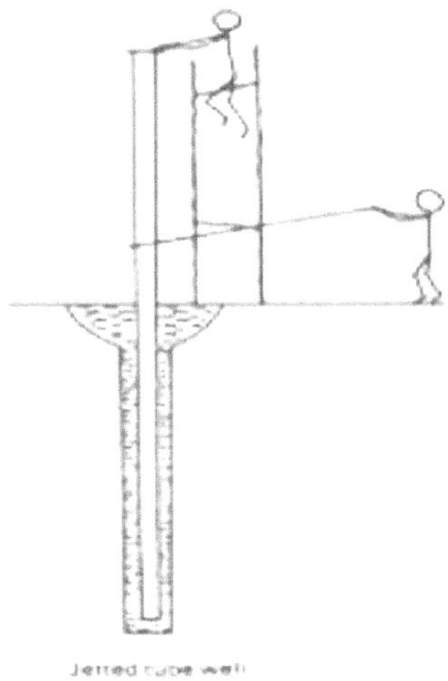

Source: SKAT (2000–2001)

- *Bored wells*
  Bored tube wells can be sunk by hand to depths up to 40m with an auger, a simple tool that is twisted to drive into the ground. Special fittings are available to pass through small rocks, remove loose soil and bail out waterlogged material from below the water table. Bored tube wells can be constructed using hand-operated equipment in soft ground; otherwise, mechanically powered equipment is necessary. The hand-operated method can be used to drill to depths of about 15 m in clay, silt and sand. The diameter of hand-drilled wells may reach 200 mm or more. With engine-powered augers, wells of large diameters (300 mm and more) may be constructed and greater depths can be reached. The yield of hand-drilled wells ranges from 5 to several hundred m^3 per day.

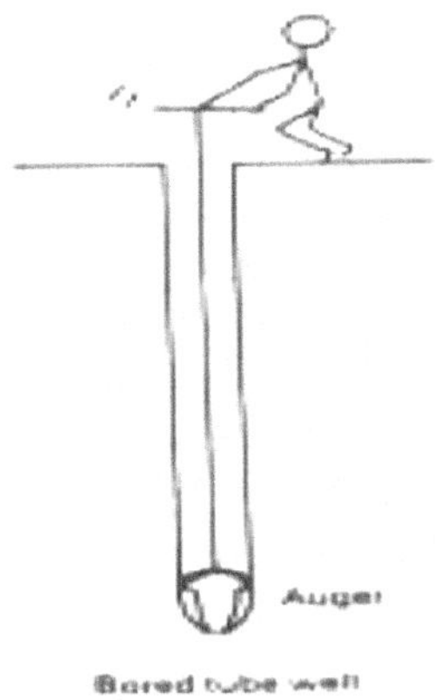

Source: SKAT (2000–2001)

- *Drilled wells*

  The term "drilled wells" is normally used to describe methods using engine-powered mechanical drilling rigs. This is the most expensive method for the construction of boreholes, and also in terms of operation and maintenance. Two methods are generally used: the percussion method and the rotary method. The percussion method employs the principle of a free-falling heavy bit delivering blows against the bottom of the well hole, forcing its way into the ground. Depending on the equipment and ground condition, wells of up to 600 mm diameter may be constructed to a depth of up to 100 m. The rotary drilling method uses a rotating bit to cut the borehole. It can be employed in any soil or rock formation. The yield of percussion wells varies from less than 100 to several 1000 m^3 per day.

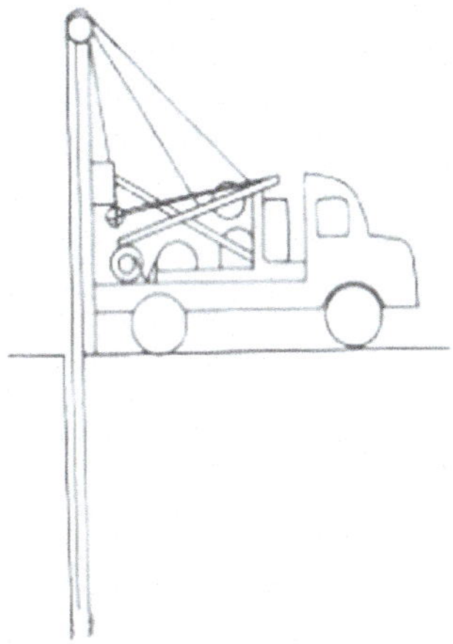

Source: WHO (1993)

I.3.3 Intakes

- *Intakes from springs*

  The main purpose of intakes from springs is to protect the springs from pollution. Two basic types of intake structure are used.
  - **Infiltration galleries or collection ditches**

    These intakes consist of perforated or open-joint pipes, sited in a gravel filter or a perforated tunnel. The infiltration gallery extends across the entire water-bearing zone to accumulate the maximum amount of water. The intake should extend below the water-bearing zone to allow the free flow into the discharge to be controlled. Anti-seepage walls may have to be constructed either along the drain ditch or close to the collection box, to prevent water from escaping. For sanitary protection, the drains or gallery should be at least 3 m below the surface. A protection zone should be established, from which any source of pollution should be excluded. Thus, there should be no housing or stables in the protection zone, and animal grazing or any other activity that might pollute the water should be prevented.
  - **Spring boxes or collection boxes**

    Spring boxes collect the water from several pipes, allowing water flow and quality to be monitored. An overflow must be established to discharge the water whenever the supply has to be disconnected. The chamber should be constructed so as to prevent contamination of the collected water. Masonry and concrete seem to be the most appropriate building materials for spring boxes.
- *Surface water intakes*

  Intakes are necessary where surface waters are used. Their placement and design depend on the type of surface water source. The intakes can be:
  - **Pipes**

    A pipe into the water is usually enclosed by a protective screen in a small frame, to keep debris from hitting and breaking the pipe and entering the system. The diameter of the pipe opening depends on the capacity, with optional velocity fluctuating between 0.5 and 1.0 m/s.

- **Infiltration systems**
  An underground infiltration system is either a well next to the water source or a perforated pipe placed in the riverbed or next to the river. The infiltration pipe is embedded in gravel and smaller rock, graded with the smallest material on the outside so that it does not obstruct the pipe. An infiltration system offers a better quality of water than that of other surface water systems because the water accumulates through a natural filter.
- **Towers**
  A tower set in the water allows water to be drawn from several different depths. Floating intakes can be also a suitable solution (see fact sheet).

River intake structures should:

- be situated in a zone with excellent water quality;
- have a constant bed and slope, with satisfactory water depth;
- be low in silt, sand and algae;
- not clog navigation or flood drainage and meet requirements with regards to dredging;
- be located as near as possible to the site where the water will be supplied;
- be positioned upstream from residential and industrial development.

I.3.4 Fact sheets

I.3.4.1 Catchment systems

- **Rooftop water harvesting**

Source: Brikké et al. (1997)

Description

The principle is to catch rainwater on the roof of a house and, using gutters and down-pipes (made of local wood, bamboo, galvanized iron or PVC), lead it to one or more storage containers. Such containers range from simple pots to large Ferro-cement tanks. A foul – flush device or detachable down-pipe is fitted to exclude the first 20 litres of run-off during a rainstorm, which is generally the most contaminated with dust, leaves, insects and bird droppings.

Operation

If there is no foul – flush device, the user has to divert the first 20 litres or so of run-off after every rainstorm. Fully automatic foul – flush devices often are not very reliable. Water is taken from the storage tank through a tap, or by pumping or using a bucket and rope.

Maintenance
Before the rainy season, the system has to be checked for holes and broken or damaged parts, and repaired if necessary. During the rainy season the system has to be checked regularly, and cleaned when dirty. This should also be done after each dry period of more than a month. Filters should be cleaned every few months. Chlorination of water may be necessary.

Advantages
Recurrent costs for materials and spare parts are very low. At small scale, this system can be operated at the community level. At medium scale, staff to run the system require no special skills. Rooftop water harvesting is useful in low- and middle-income countries with substantial rainy seasons and where other improved water supply systems are difficult to realize.

Disadvantages
During certain periods of the year, water quantity may be insufficient for domestic purposes and drinking-water quality standards may not be met. The investment needed for the construction of a tank and suitable roofing may be beyond the financial capacity of poor households. Acceptability will depend on the taste of the water.

- **Catchment areas and storage dams**

Source: Brikké et al. (1997)

Description
Through the construction of a dam, the rainwater or snow catchment areas of a natural surface such as a bedrock area or a valley can be made available for human use. Water is stored behind the dam or diverted to a separate reservoir. Important factors in the planning of dams are annual rainfall, evaporation patterns, present use, the run-off coefficient of catchment areas, water demand, the geology and geography of the area, and the site where the dam will be built. Dams (<3m in height) may consist of raised banks of compacted earth (usually with an impermeable clay core, store aprons and a spillway to discharge excess run-off), masonry or concrete (reinforced or not). Users can take water directly from the reservoir or water can be treated and distributed through a larger system.

Operation
Operational activities by a caretaker may include opening and closing valves or sluices in the dam or in the conduits to a reservoir. Water is usually extracted from the water points by users.

Maintenance

Throughout the year, animals have to be kept away from the catchment area and reservoir. Contamination with human waste must be avoided. The dam, valves, sluices and conduits have to be checked for leaks and structural failures. The catchment area must be checked for contamination and erosion. To avoid mosquito breeding and the possible spread of malaria, *Tilapia* or other larvivorous fish can be introduced in the reservoir. Reservoir management approaches can also be applied for the control of mosquitoes and of snail intermediate hosts of schistosomiasis (bilharzia).

Advantages

Storage dams provide a water source available for continuous use. Recurrent material costs are usually low. A caretaker (possibly a person who lives or farms near the site) should be appointed.

Disadvantages

Contamination of the water is a potential problem. Also, there are risks of water-borne and water-related diseases such as schistosomiasis and malaria. Even if the law does provide some protection, catchment areas are vulnerable to damage by animals and people.

- ## Spring water collection

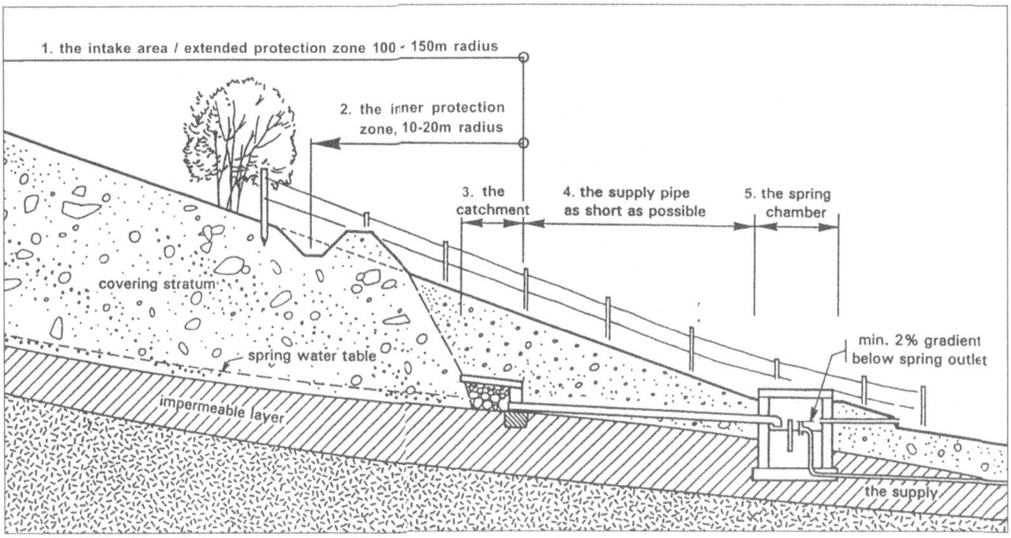

Source: SKAT (2000–2001)

Description

A spring-water collection system catches and protects ground water flows at the points where these arrive at the surface. The main parts of a spring-water collection system are a drain under the lowest natural water level, a protective structure providing stability, and a seal to prevent surface water from leaking in. The drain is usually placed in a gravel pack covered with sand, and may lead to a conduit or a reservoir. To prevent contamination through infiltration from the surface, a ditch, known as the interceptor drain, diverts surface water away from the spring box. A fence keeps animals out of the area around the spring.

Operation

Water should be permitted to flow out freely all the time to prevent it finding another way out of the aquifer. Operational activities may include opening or closing valves to divert the water to a reservoir or a drain. The spring and its surroundings must be kept clean.

Maintenance

The purpose of maintenance is to prevent pollution, both in the area where the spring water infiltrates into the ground and in the immediate surroundings of the spring. The water flow from the spring box must be checked. If there is an increase in turbidity, surface run-off has to be identified and the protection of the spring improved. If water flow decreases, the possibility must be considered that the collection system is clogged.

Advantages

Recurrent material costs are usually very low. Caretakers will need to be employed, but their cost will usually be low as well, as usually, is the cost of the water conveyance system.

Disadvantages

Problems may arise if a large investment is suddenly needed for construction or for a major repair or for the replacement of the system. Some springs may not deliver enough water throughout the year. Not all springs produce clean water of an acceptable taste. Springs may be sited too far from the households or on privately owned land.

I.3.4.2 Wells

- **Dug wells**

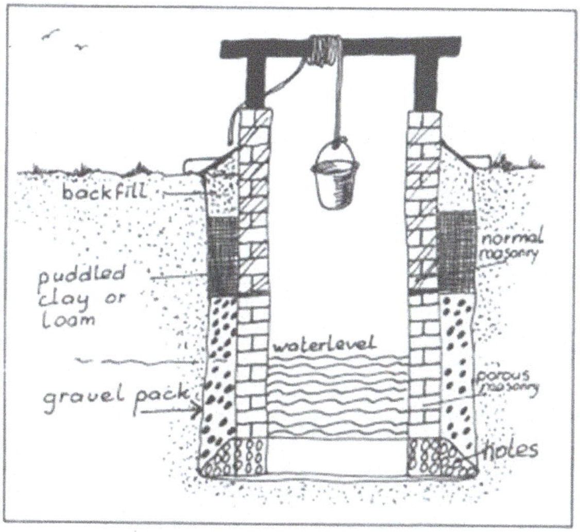

Source: Brikké et al. (1997)

Description

A well gives access to groundwater from an aquifer and facilitates its extraction. A dug well in particular refers to a well that is large enough to enable a person to get into it to clean or deepen it.

A dug well will therefore rarely be less than 0.8 m in diameter. Its structure consists of three main components: a stone, brick or concrete apron; the well lining between ground level and water level; and the well lining below water level. Other components often found are: a drain; and a fence around the well, with a gate.

Operation
If water is abstracted with a rope and bucket, the operational activity may consist of taking off and putting back the cover of the well.

Maintenance
In general, only light maintenance is required. There should be daily checks for any debris visible in the well, and such debris should be removed. The concrete apron will need to be cleaned. The fence and drainage should be checked, and repaired or cleaned as necessary. If the well has gone dry or does not yield enough water, it has to be deepened and lined further down. The users of the system or a caretaker can generally carry out the light maintenance that is required. Major repairs will call for skilled labour, which will usually be available locally.

Advantages
Recurrent material costs are usually low. Caretakers will need to be assigned, but their costs will usually be low as well. Dug wells can often be constructed with locally available tools, materials and skills. If the water lifting system breaks down and cannot be repaired, the well can continue to be used with a rope and bucket. A dug well has substantial storage capacity. Also, it can be constructed in geological formations where hand-drilling or even mechanical drilling is difficult or impossible.

Disadvantages
Wells constructed far from users' homes or in places that are difficult to reach will not be used much, and are likely to be inadequately maintained. Wells in such locations will also incur considerable opportunity costs such as irrigation. Wells should not be sunk near latrines.

- **Drilled wells**

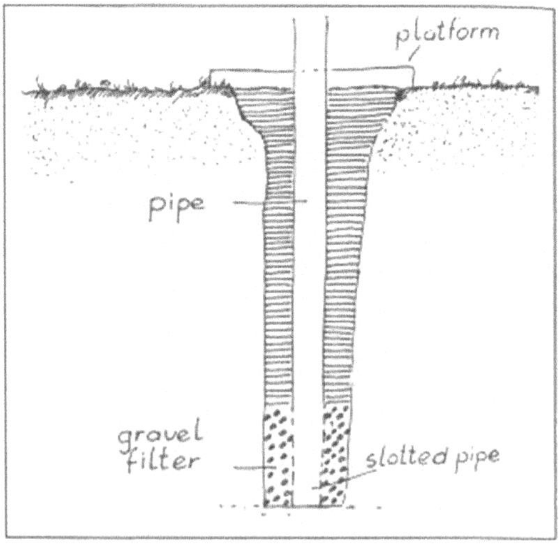

Source: Brikké et al. (1997)

Description
Drilled wells, tube wells or boreholes differ from dug wells in that they are much narrower, generally varying between 0.1 m and 0.25 m in diameter. A person cannot enter a drilled well to clean or deepen it.

Operation
No operational activities are generally needed for the well itself. If the production capacity of the well is lower than demand, however, daily monitoring of the water level may be necessary. Water is usually extracted from the well by users or by a caretaker.

Maintenance
Apart from cleaning the apron daily, and occasionally cleaning the drain and repairing the fence (if these are present), there are hardly any maintenance activities or related costs. If the well becomes clogged, various technologies may be used to clean it, such as forced air and water pumping, brushing, and treatment with chemicals. It is very difficult to deepen an existing drilled well.

Advantages
Recurrent material costs are usually low. Caretakers will need to be assigned, but their costs will usually be low as well. Water for irrigation may be an added beneficial use of the well.

Disadvantages
Occasionally, major maintenance activities such as rehabilitation of the well may suddenly require a high investment. This may pose problems if the investment has to be financed by the community. The life expectancy of a good well is more than 20 years, but after a few years the yield may diminish drastically and rehabilitation may be necessary. Well construction depends on geo-hydrological conditions, depth and yield of aquifers, and the presence of rock formations above them. Wells constructed far from users' homes or in places that are difficult to reach will not be used much, and are likely to be inadequately maintained. Wells in such locations will also incur considerable opportunity costs such as irrigation. Wells should not be sunk near latrines.

- **Subsurface harvesting systems**

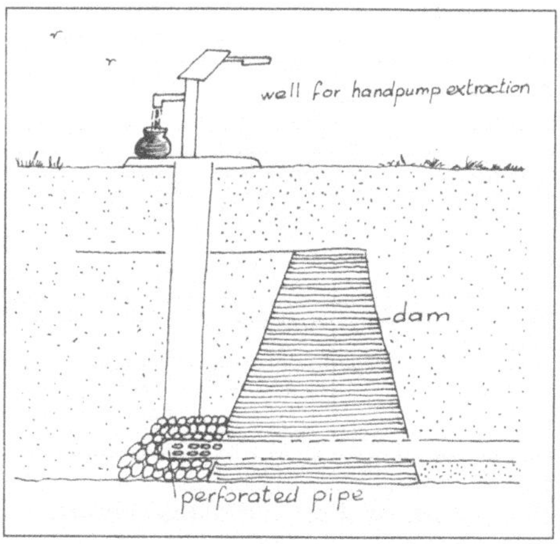

Source: Brikké et al. (1997)

Description
Subsurface harvesting systems retain groundwater flows and facilitate their extraction. There are two main types: *subsurface dams*, where an impermeable dam is built across a surface aquifer, with its base on an impermeable layer and its crest about 1 m under the level of the surface (to prevent land becoming waterlogged); and *raised-sand dams*, where an impermeable dam is built across the bed of a seasonal sand-filled river, with the crest of the dam a few decimetres higher than the upstream riverbed.

Both dam types have wing walls embedded in the riverbanks against which rocks may be piled up to prevent erosion.

Operation
Operational activities may include opening or closing the gravity pipe, control and monitoring tasks, and operation of the water supply systems.

Maintenance
Only light maintenance is generally required. This includes regular cleaning of the well or gravity pipe, as necessary. After every flood, the dam needs to be checked to see if the water has damaged it. During the dry season, the crest of a raised-sand dam should be raised by a maximum of 50 cm if the reservoir has filled up.

Advantages
For subsurface dams, recurrent material costs are usually low. Caretakers will need to be assigned, but their costs will usually be low as well. Raising a crest of a dam or make repairing after flood damages may involve higher costs.

Disadvantages
Possible construction sites may be far from the homes. Substantial opportunity costs may be incurred. Locally available materials may not meet the standards for construction. The investment in labour, in cash or in kind, needed for the construction of subsurface harvesting systems may be beyond the capacity of low-income communities. Subsurface harvesting is inappropriate where the resulting rise of the water table could have a negative impact on agriculture, infrastructure works or buildings.

I.3.4.3 Intakes

- **Protected side-intake**

Source: Brikké et al. (1997)

Description

A protected side-intake provides a stable place in the bank of a river or lake, from which water can flow into a channel or enter the suction pipe of a pump. Side intakes are sturdy structures, usually made of reinforced concrete, and may have valves or sluices to flush out any sediment that might settle. Often a protected side-intake is combined with a weir in the river (which keeps the water at the required level), a sand trap to let sand settle, and a spillway to release excess water.

The river water may enter the side intake through a screen, and a spillway overflow may be provided. Sometimes protected side-intakes are combined with a dam and a flushing sluice that enables the upstream part of the river to flush away.

Operation

Protected side-intakes are usually operated by a caretaker. A valve or sluice may have to be adjusted on a day-to-day basis. The inlet to the channel or pump must be checked daily, and any debris obstructing it must be removed and any damage repaired.

Maintenance

Preventive maintenance consists of painting screens and other metal parts, such as sluices or valves. Depending on the level of sediment, the intake canal and silt trap have to be desilted. Any debris has to be regularly cleaned from screens, and if screens are damaged, they have to be welded. During the rainy season the inlet may have to be checked and cleaned more frequently. Any erosion damage to the riverbank or bed has to be repaired immediately with boulders, rocks, bags filled with sand, and so on Cracks in the concrete structure have to be repaired every year.

Advantages

In general, recurrent costs are low, especially if the inlet is not subject to erosion.

Disadvantages

River and lake water may be polluted. Clogging by silt or debris may occur. River currents may undermine the infrastructure.

- **River-bottom intake**

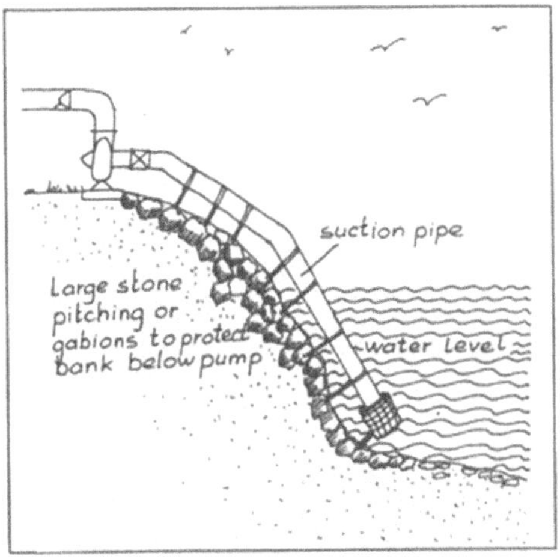

Source: Brikké et al. (1997)

Description
River-bottom intakes for drinking-water systems are usually applied in small rivers and streams where sediment content is low. The water is extracted through a screen over a pipe inlet (the pipe is usually made of concrete) that is built into the riverbed. The bars of the screen are laid in the direction of the current and slope downwards so that coarse material cannot enter the pipe but continues its course downstream. From the canal, the water enters a sand trap. After that, it may pass through a valve and flow by gravity or be pumped into the distribution system.

Operation
River-bottom intakes are usually operated by a caretaker. The inlet must be checked regularly. Debris obstructing the inlet must be removed and any damage repaired. The sand trap must be cleaned regularly.

Maintenance
Preventive maintenance consists of painting screens and other metal parts, such as sluices or valves. The sand trap and screen will have to be cleaned regularly, with the frequency of cleaning depending on the silt and bed-load transport. Sometimes the screen or a valve may need repair. Any erosion undermining the structure must be repaired immediately. Each year the concrete structure has to be checked for cracks and, if necessary, repaired.

Advantages
Recurrent costs are low, especially if the inlet is not subject to erosion.

Disadvantages
River and lake water may be polluted. During the dry season, the amount of available water may be too low. Clogging by silt or debris may occur. River currents may undermine the infrastructure.

- **Floating intake**

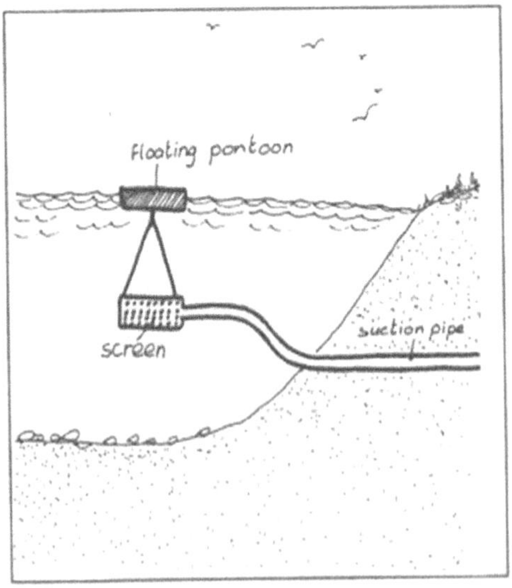

Source: Brikké et al. (1997)

Description

Floating intakes for drinking-water systems allow extraction of water from a river or lake near the surface, thus avoiding the heavier silt loads transported close to the bottom during floods. The inlet of the suction pipe of a pump is connected just under the water level to a floating pontoon that is moored to the bank or bottom of the lake or river. The pump itself can be located on the bank or the pontoon. The advantage of placing the pump in the pontoon is that it can be combined with a turbine to use the energy from the water current and that the suction pipe can be extremely short.

For the construction of the pontoon, a steel or wooden frame can be made, with floaters of empty oil drums, plastic containers or sealed steel tubes of at least 30 cm diameter.

Operation

Floating intakes are usually operated by a caretaker. The pump inlet must be checked before and during pump operation; any debris obstructing it must be removed and any damage repaired. Mooring cables have to be checked daily and adjusted if necessary.

Maintenance

Flexible pipe connections must be checked daily for leaks, and the pump inlet must be inspected daily. If the inlet gets obstructed, it must be cleaned. Any damage to the mooring or the pontoon structure must be repaired immediately. The pontoon has to be painted regularly. The frequency of painting will depend on the materials used, but should be at least once a year for steel parts.

Advantages

Limited expenses may be expected for welding and painting, replacement of cables and some other material costs.

Disadvantages

Collision with floating objects is a risk. Pipe connectors between pontoon and bank may become worn from time to time, and may need to be replaced. Lakes and rivers may be of poor water quality.

- **Sump intake**

Source: Brikké et al. (1997)

Description

In a sump intake, water from a river or a lake flows through an underwater pipe to a well or sump from where it is lifted, usually into the initial purification stages of a drinking-water system. The inflow opening of the underwater pipe is located below the lowest water level and is screened. The well provides a place for sedimentation and protects the pump against damage by floating objects. Sometimes two sump intakes are built for one pump in order to facilitate cleaning.

Operation

Sump intakes are commonly operated by a caretaker. The sump must be checked daily for sufficient water inflow; any debris obstructing it must be removed, and any damage repaired.

Maintenance

Most of the maintenance concerns the pump. The intake itself needs some cleaning and desilting. If it caves in or if the river or lake bank erodes, repairs have to be made.

Advantages

Apart from some labour costs, sump inlets present few recurrent costs.

Disadvantages

Silt or debris may clog the inlet pipe. Erosion caused by river current may undermine the intake structure and the bank. Water quality from rivers and lakes may be poor. Sump intakes are not suited for rivers that are very shallow and have low flow rates.

I.4 WATER CONVEYANCE TECHNOLOGIES

Water conveyance is associated with a considerable amount of additional operation and maintenance activities, and with the risk of potential problems, so gravity systems are generally preferred, where available. However, if distances to a gravity water source are very long, or if the quality and quantity of water are inadequate, water will have to be lifted.

For this purpose, a great variety of conveyance technology options exists, from the simple manual rope and bucket to more sophisticated types of pumps.

I.4.1 Pumps

Pumps are complex machines; it is therefore recommended that their installation, operation and maintenance should be undertaken by trained and experienced staff.

I.4.1.1 Classification of pumps

There are two basic types of pumps: kinetic or positive displacement. Kinetic pumps can be centrifugal, peripheral or special (such as gas, jet, hydraulic ram and electromagnetic pumps). Positive displacement pumps are basically rotary and reciprocating (hand pumps).

The most commonly used among the kinetic pumps is the centrifugal pump, which may be single-stage or multistage. The centrifugal pump, in its various forms, has almost completely displaced all other types of pumping equipment for piped water supplies. The hand pump uses human and animal power. Other types of pumps include rotary, injection or jet, air-lift, and hydraulic ram pumps, and are largely used for special applications.

For small communities in low- and middle-income countries, human or animal power is often the most readily available for pumping water, particularly in rural areas. Under suitable conditions, wind power may be an option. Solar energy has potential but has been little used for water pumping to date. Diesel engines and electric motors imply much greater expense, both in capital and recurrent costs, and also require skilled operation and maintenance. They should only be used if the necessary energy (fuel or electricity) and other operating costs are affordable, if adequate trained personnel for maintenance are present and if spare parts can be obtained.

I.4.1.2 Selection of pumps

For the selection of pumps, the following considerations are important.

- **Nature of the water to be pumped**
  The nature of the water will determine the type of impeller best suited for the purpose and the material to be used for pump construction. For pumping clear water, pumps with single or double suction and close-type impellers are recommended. For pumping muddy water, pumps with mixed flow impellers are most suitable. For pumping brackish water, corrosion-resistant materials should be used.
- **Pumping capacity and lifting head**
  For small units, the single suction, closed impeller type of pump is more practical. For larger plants, the horizontal shaft, double suction, volume casing centrifugal pump is the most commonly used type.
- **Initial cost of pump and its driving equipment**
  Efficiency is a very important factor where electric engines are used, particularly where power is expensive. In some circumstances, a new pump installed to replace an old, inefficient one would pay for itself within a very short time simply by reducing electricity consumption.
- **Extent of maintenance, and reliability**
  Pumps in which all moving parts are above ground and easily accessible are preferable, and will in most instances be easier to maintain. However, if qualified maintenance can be assured, pumps with submerged cylinders should be considered.
- **Desirability of standardization and reduction of spare parts**
  When a number of pumps are involved, it is desirable to standardize the units so that parts are easily interchangeable in times of emergency, and to reduce the number and diversity of spare parts to be stocked for repair and maintenance purposes.
- **Fluctuations in suction levels**
  Suction lift is of great importance when dealing with centrifugal pumps. Normally, the pump should be installed such that the distance of the centre line of its suction opening from the lowest water level obtained in the well when pumping will not exceed 4.5 m. If the suction lift exceeds the maximum value, serious problems such as cavitations may result.

I.4.2 Power systems

Pumps for water supply systems are normally powered by hand, electric motors, or diesel or petrol engines. In specific cases, pumps may be powered by animals, water (turbine or hydraulic ram), windmills or solar cells.

- *Human power*
  The use of human power for pumping is restricted to small systems. Human pumping rates commonly do not exceed about 30 l/min at lift heads of several metres only, and may become

much lower for larger heads. Experience shows that hand pumps used by the general public, as contrasted with hand pumps for individual households, exhibit excessive problems in regard to wear and maintenance.

- *Electric power sources*
  Where electricity is available from a central supply nearby, it is the most convenient and economical power source for a pump. Small electric engines are easy to operate, and their purchase and maintenance costs are low. The electric power source should be reliable and not subject to significant voltage variation. If the reliability is low and power outages of several hours are frequent, sufficient elevated storage or an emergency power source should be provided. The required power supply in KW can be estimated from pump tables supplied by pump manufacturers.
- *Combustion engines*
  In the absence of electricity or as a standby in the event of power failure, diesel, petrol or gas engines may be used to drive pumps directly. Normally this arrangement is cheaper and more efficient than using a motor-driven generator to produce electricity to drive the pump. However, where submersible pumps are used, an electricity generator driven by a combustion engine will be necessary.
- *Other power sources*
  In certain situations, other power sources – such as windmills, water turbines and solar cells – may become technically and economically feasible.

I.4.3 Fact sheets

I.4.3.1 Pumps and lifting devices

- **Rope and bucket through a pulley or on a windlass**

Source: Brikké et al. (1997)

Description

Rope and bucket systems are mostly used with hand-dug wells. A bucket on a rope is lowered into the water. When hitting the water, the bucket dips and fills itself. The bucket is then pulled up with the rope. For well depths of less than 10 m, a windlass can be used, with a hose running from the bottom of the bucket to a spout at the side of the well. Even with this system and a protected well, hygiene is poorer than with a bucket pump.

Operation
The bucket is lowered or raised by paying out or pulling in the rope, or rotating the windlass. Care must be taken not to dirty the rope or bucket.

Maintenance
Preventive maintenance consists of greasing the bearings of the windlass or pulley. Small repairs are limited to the patching of holes in the bucket and hose, reconnecting the bucket hinge, and fixing windlass bearings or handle. All repairs can be done locally, using tools and materials available in the local community.

Advantages
Depending on the quality of materials, there is usually only a need for occasional purchases of rope, bucket, hose, wire, and so on Also, occasional repair costs of the windlass are low. This is a global option, mainly in rural areas.

Disadvantages
Bad quality rope deteriorates fast. There is a risk of the bucket falling into the well. In systems using a windlass with hose, the hose may break frequently. Hygiene is in jeopardy, especially when the rope and bucket touch people's hands and the ground. Communal wells often tend to get more contaminated than family-owned wells.

- **Bucket pump**

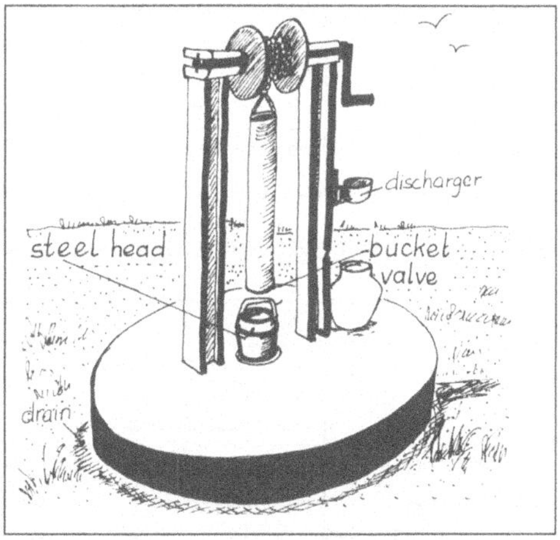

Source: Brikké et al. (1997)

Description
Bucket pumps are mostly used in drilled wells. They consist of a windlass over a 125 mm PVC tube, down which a narrow bucket with a valve at the bottom is lowered on a chain. When the bucket hits the water, the valve opens and water flows in. When the bucket is raised, the valve closes, keeping the water inside. To release the water, the operator of the pump rests the bucket on a water discharge, which opens the valve at the bottom. The bearings of the windlass are made of wood.

Operation
Operating a bucket pump involves rotating the handle of the windlass and manoeuvring the bucket. Adults and children can operate the pump.

Maintenance
Maintenance involves keeping the pump and its environment clean. Preventive maintenance consists of lubricating the wooden bearings of the windlass, checking nuts and bolts, and checking the functioning of the valve. Regular small repairs include the replacement of the valve washer, and the repair of a link in the chain. A main repair is the fixing of the bottom edge of the bucket. The chain, the bucket or the bearings of the windlass will need replacement at times.

Advantages
A bucket pump is easy for an operator to use. Recurrent costs are mainly linked to the purchase of new washers for the valve.

Disadvantages
Bucket pumps have low discharge rates. Communal wells risk contamination, and chlorine to disinfect wells is often not locally available.

- **Rope pump**

Source: Brikké et al. (1997)

Description
The basic parts of a rope pump are a pulley wheel above the well, a riser pipe from under the water level to an outlet just under the wheel, and a rope with rubber or plastic washers that goes through the pipe over the wheel back down into the wheel and into the pipe again. When the wheel is being turned, the upward moving washers lift water into the pipe towards the outflow. The rope pump can be made completely at village level using wood, rope, and PVC or bored-out bamboo tubing.

Operation
Men, women and children can operate a rope pump. Turning the handle of the pulley wheel draws the water up. After pumping, the wheel has to be held for a moment in order to allow the water in the riser pipe to drain, and to prevent the washers from being pulled back in the pipe, causing extra wear. The site and the pump must be kept clean.

Maintenance
Depending on the intensity of use of the pump and the type of bearings, the axle bearings must be greased at intervals of not more than one week. Fixations of the pulley wheel and the rope have to be checked regularly. Between about every six months and three years, the rope has to be replaced. Every few years, the washers have to be renewed. Tubes last for at least six years; the frame and pulley wheel of the pump can last from six to twelve years; and the rope guide should last for several years. All repairs can be done by users, sometimes with the assistance of a skilled worker for welding.

Advantages
Although the quality of design and construction may differ significantly, the rope pump has the potential to be low in cost, and to be operated and maintained at the village level.

Disadvantages
Traditional rope pumps have a lift of only about 10 m. The pump requires more care by users than many other pumps, and is more susceptible to contamination. Excessive wear of the rope may result from exposure to the sun. Rope exposed to the open air needs to be protected. There is a risk of poor installation of an inferior quality pulley or guide block. Problems may arise with the pulley wheel. The pistons and frame structure may be of poor quality.

- **Suction plunger hand-pump**

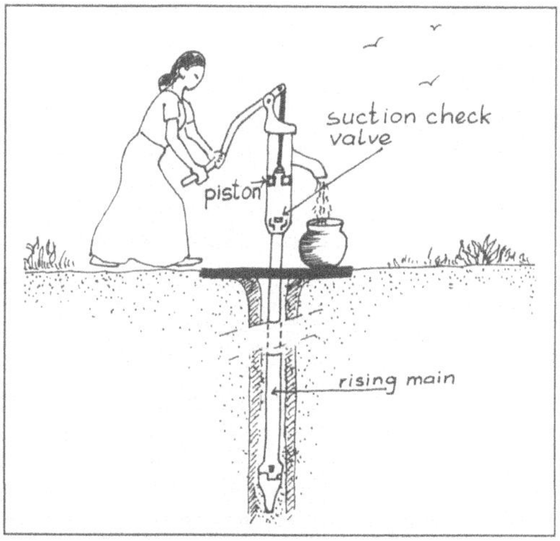

Source: Brikké et al. (1997)

Description
A suction plunger hand-pump has its cylinder and plunger (or piston) located above the water level, usually within the pump stand itself. These pumps must be primed by pouring water on the plunger.

On the upstroke of the plunger, the pressure in the suction pipe is reduced and the atmospheric pressure on the water outside pushes the water up into this pipe. On the down stroke, a check valve at the inlet of the suction pipe closes and water passes the plunger through the opened plunger valve.

With the next upstroke this water is pushed up by the plunger and flows out at the top, while new water flows up in the suction pipe.

Operation

Operation starts with priming the pump by pouring clean water on the plunger through the top of the pump stand. Pumping involves moving a handle up and down, usually while standing beside the pump.

Maintenance

Suction pumps are relatively easy to maintain. This can be done by a village pump caretaker or by users themselves, using simple tools, and basic spare parts and material. Preventive maintenance consists of greasing bearings and inspection of the interior of the pump stand. During these inspections, smaller repairs such as replacement of washers may prove to be necessary. For major repairs, more highly skilled workers as well as more specialized tools and material may be needed.

Advantages

Suction plunger hand-pumps are easy to use in rural and low-income areas where groundwater tables are within 7 m from the surface. The pump may function without major problems for more than seven years.

Disadvantages

The maximum pumping lift of about 7 m is the biggest drawback of suction pumps. Suction pumps must be primed, for which contaminated water is often used. Most pumps are designed for family use and are not sturdy enough for more intensive communal use.

- **Deep well diaphragm-pump**

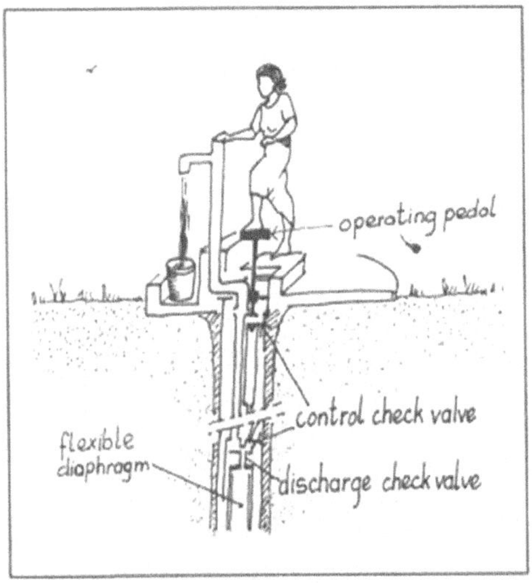

Source: Brikké et al. (1997)

Description

Inside a cylindrical pump body at the bottom of the well, a flexible diaphragm shrinks and expands like a tube-shaped balloon, taking water in through an inlet valve connected to a flexible hose, which leads the water to the surface. The piston is usually moved by depressing a foot pedal. When foot pressure is removed, the elasticity of the diaphragm forces water out and back up the pilot pipe to lift the pedal. The piston hand-pump is another similar type of pump.

Operation
The pump is operated by pushing down the pedal by foot or using a handle. Considerable effort is needed to push the pedal; this is acceptable, as full body weight can be applied.

Maintenance
Every day, the pump head, platform and surroundings should be cleaned, and nuts and bolts should be tightened. Every month, the drive piston, the rings and guide bushing should be checked and replaced if necessary. Depending on borehole conditions and at least once a year, the down-hole parts have to be checked and the whole pump should be washed with clean water. A major repair is the replacement of the diaphragm.

Advantages
The principle of the pump is attractive because it allows the use of thin flexible hoses, making it easy to install or remove without the need for special tools or materials. Several pumps can be installed in a single well or borehole.

Disadvantages
Deep well diaphragm-pumps are unsuitable for use by children and pregnant women due to the high effort required to operate the pedal. If the community is suddenly unable to pay the high costs for replacement of the diaphragm or if no skilled mechanic is available when such replacement is needed, users may temporally be forced to return to their traditional sources.

- **Centrifugal pump**

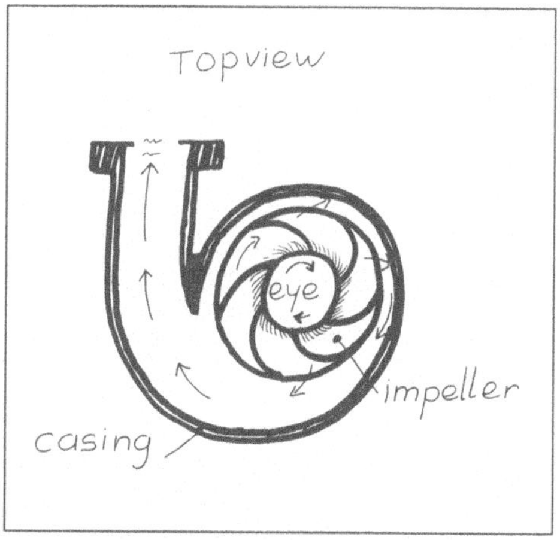

Source: Brikké et al. (1997)

Description
The essential components of a centrifugal pump are the fast rotating impeller and the casing. Water flows in at the centre of the impeller, which imparts kinetic energy to it, producing an outward flow due to the centrifugal forces. The kinetic energy of the water is partly converted to useful pressure which forces the water into the delivery pipe. The water leaving the eye of the impeller creates suction, which draws water from the source into the pump. In order to be able to place the pump above the suction limit, some pumps inject a jet stream of water at the inlet of the suction pipe.

Operation

During pumping, operation involves checking the condition of the engine, the water output of the pump, and any vibration. If the pump house runs dry, clean water has to be poured into it. The pump inlet has to be maintained, and the pump and engine kept clean. Running hours, problems, servicing, maintenance and repairs should be recorded in a logbook.

Maintenance

Dismantle the pump annually. Remove the rising main from the well and inspect it. Check the inlet screen, foot valve and pipe threads, and re-cut corroded or damaged threads. The foot valve may need a new rubber washer or may have to be replaced. All other repairs, such as the replacement of bearings or the impeller, involve high costs and have to be done by qualified technicians.

Advantages

This is an improved type of pump, for which no human power is needed. The costs of spare parts, materials, tools and equipment are generally low compared to the costs for fuel or electricity.

Disadvantages

The price electricity or fuel, the reliability of the electricity or fuel supply, and the need for skilled technicians are all disadvantages. Debris, sand or other particles entering into the pump may cause abrasion damage. Cavitations damage may be caused by a clogged inlet. The pipeline system may be damaged by severe pressure surges caused by abrupt starting and stopping of the pump. Fast wear of bearings will occur if the pump and engine are badly aligned.

- **Submersible pump**

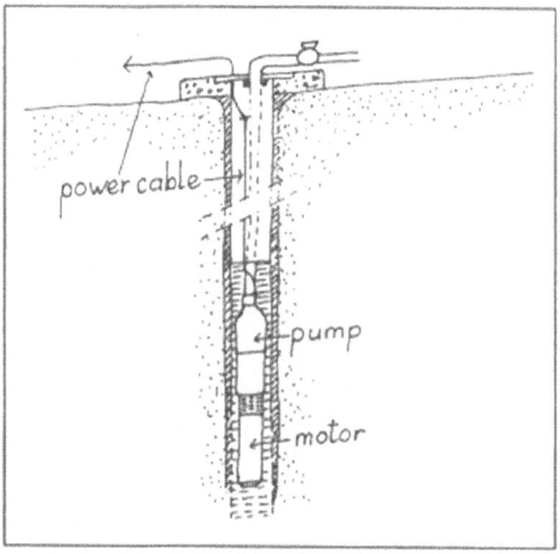

Source: Brikké et al. (1997)

Description

Because both the pump and engine are located underwater, this is called a submersible pump. Usually this pump is placed above the engine and under a check valve that leads to the rising main. In order to prevent the pump from running dry, the water level in the well must be monitored, and pumping must be stopped if the water level drops to the intake of the pump. Power is delivered through a heavily insulated electricity cable connected to a switch panel at the side of the well.

Operation

During pumping, operation involves checking the clearness of the water flow and the power consumption of the pump. If the water is turbid, the well has to be cleaned or else the pump will wear quickly. Running hours, problems, servicing, maintenance and repairs should be recorded in a logbook.

Maintenance

Remove the pump and the rising main from the well and inspect them annually. Check the inlet screen, the valve and pipe threads, and re-cut corroded or damaged threads. Replace badly corroded pipes. Inspect electrical cables and check the insulation between cables. All other repairs, such as replacement of the pump, involve high cost and have to be done by a skilled technician.

Advantages

This is an improved type of pump, for which no human power is needed. The costs for spare parts, materials, tools and equipment are generally low compared to expenses for electricity (depending on the country).

Disadvantages

The price and reliability of supply of electricity or fuel, and the high technology level are the main limitations. Sand or other particles entering the pump may cause abrasion damage. There is also a risk of corrosion of the rising main. Damage may be caused by abrupt starting and stopping of the pump.

- **Axial flow pump**

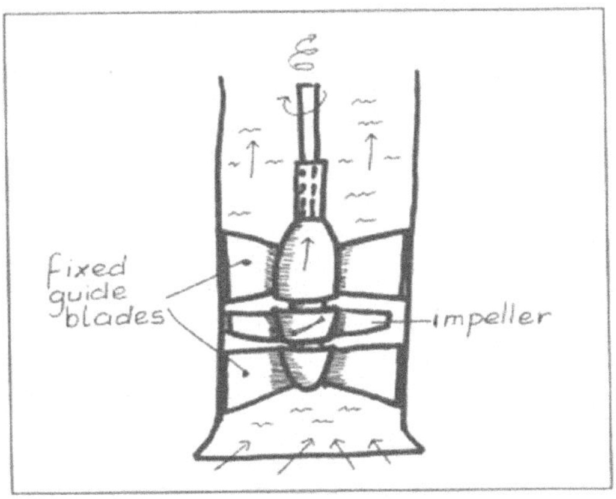

Source: Brikké et al. (1997)

Description

The axial-flow or propeller pump is similar to a ship's propeller in the way it moves the water; in fact, a boat propeller is often used to improvise an axial flow pump. In contrast to the centrifugal pump, the water flows through the pump in the same direction as the drive shaft. These pumps are specifically suited for high flows and low-pressure heads. The axial flow pump turns in a cylindrical casing, where fixed guide blades may be installed in order to minimize whirls.

Many rotary pumps use an impeller that is shaped somewhere between an axial-flow and a centrifugal impeller, to serve specific situations.

Operation

Before operation, the caretaker has to check that the inlet is clean. During operation, the caretaker has to check that the pump is running smoothly and that water output is normal. Other operational activities include maintaining the pump inlet, and keeping the pump and engine clean. Running hours, problems, servicing, maintenance and repairs should be recorded in a logbook.

Maintenance

Dismantle the pump annually and check the impeller blades, guide vanes and bearings for damage; also check them for oxidation. Repair, paint or replace them if necessary. Grease the bearings, as appropriate. Replacement of bearings, seals or impeller sometimes involves high cost and has to be done by qualified technicians.

Advantages

Only low pumping heads are needed.

Disadvantages

Debris, sand or other particles entering the pump may cause abrasion or other damage. Cavitations damage may be caused by a clogged inlet. Damage to the pipeline system as a result of severe pressure surges may be caused by abrupt starting and stopping of the pump. Rapid wear of bearings may occur if the pump and the engine are badly aligned. The price and reliability of supply of electricity or fuel, and the high technology level of the pump are the main limitations.

- **Hydraulic ram pump**

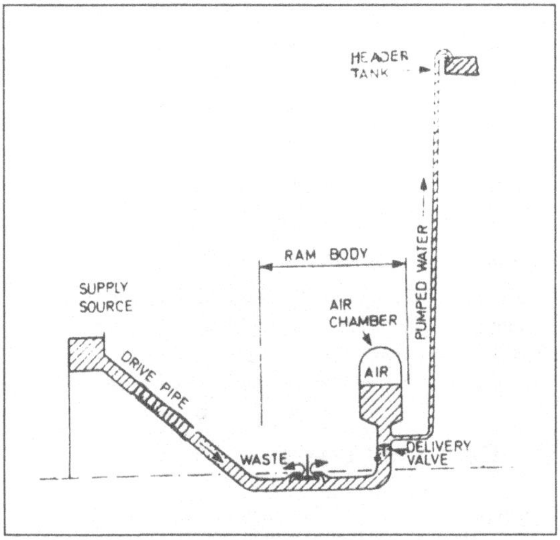

Source: Brikké et al. (1997)

Description

Hydraulic rams are devices that pump water using the shock energy of a flowing water mass that is suddenly forced to stop. Only part of the water used is pumped up to a higher altitude. At the beginning of the pump cycle, water starts to flow from the source down through an inclined rigid drive pipe several metres long, into the hydraulic ram body and then out through the impulse valve. If the water has accumulated enough velocity, it forces the impulse valve to close with a blow, and the water in the drive pipe suddenly comes to a stop. This produces a shockwave in the water mass that forces an amount of water through the delivery valve, which is located in the pump body, into a buffering air chamber and from there up the delivery pipe.

Other designs use a buffer chamber that contains a piece of compressible rubber instead of air.

Operation

Hydraulic rams have to be started by hand, repeatedly opening the impulse valve until the pump continues to operate by itself. The weight or spring tension on the impulse valve may have to be adjusted to reach the right frequency. The pump inlet, the pump and the site must be kept clean.

Maintenance

Proper functioning of the delivery valve must be checked weekly, and bolts should be tightened. Occasionally, the pump will have to be dismantled and cleaned to remove the sand and silt accumulated in it. Once in a while, the rubber washers of the valves will have to be replaced, the frequency of replacement depending greatly on the rubber quality.

Advantages

Because of the low cost of a hydraulic ram pump and intake works, this technology is suitable for communal use.

Disadvantages

Hydraulic rams need a water supply that is at least one metre higher than the pump body. When the pumping lift is high, the amount of water pumped is only a fraction of the intake through the drive pipe. For that reason, output flow is generally low.

I.4.3.2 Power systems

- **Electric power sources**

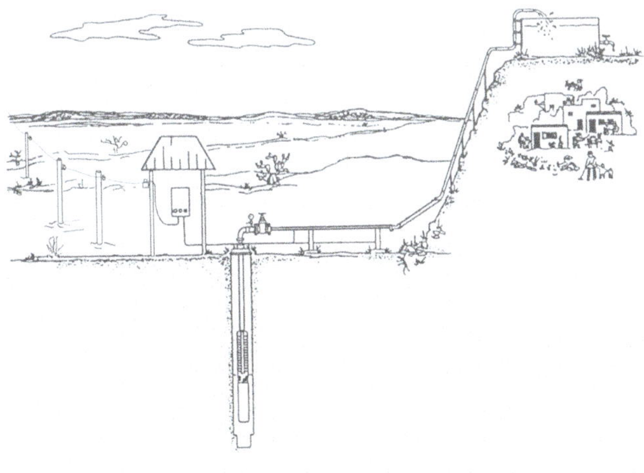

Source: SKAT (2000–2001)

Description

Where electricity is available from a central supply system nearby, it is the most convenient and economical power source for a pump. The costs of these systems vary depending on the distance of the pump site from the electric grid, the size of the distribution network, and the type of reservoir.

Operation
The electric power source should be reliable and not subject to significant voltage variation.

Maintenance
Submersible electric pumps can operate with little maintenance. Repairs mainly concern the stop-to-start pump control device.

Advantages
If a distribution network with standpipes is already installed, then the access to the source is relatively easy.

Disadvantages
A water storage tank is required to ensure water supply during the period when power supplies are suspended and to balance the hourly fluctuations in demand.

- **Diesel engine**

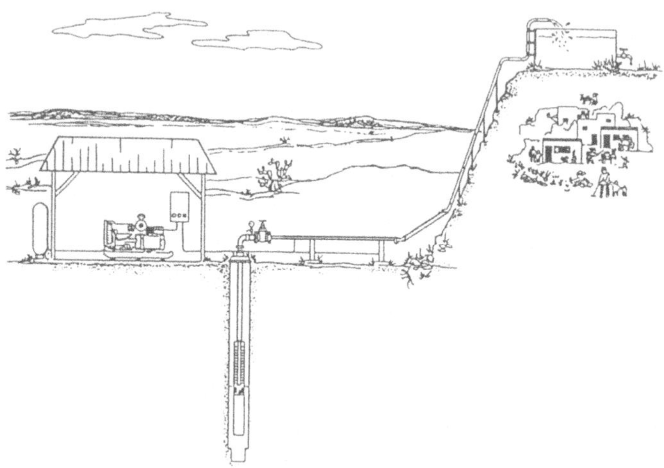

Source: SKAT (2000–2001)

Description
Diesel engines are very often used as a stationary power source. The main parts of the engine are the cylinders, pistons, valves and crankshaft. Diesel engines are more efficient than petrol engines, and they need less maintenance. Diesel engines differ in size, speed, cycle, cooling system, etc.

Operation
A diesel engine must be operated by a trained caretaker. There are specific operating instructions for each type of engine. Before starting the engine, the fuel, oil and cooling-water levels should be checked.

Maintenance
Every day, the outside of the engine must be cleaned. In dusty conditions, the air filter must be checked and cleaned. Some parts may need manual lubrication. The engine is serviced for preventive maintenance according to the number of hours it has run.

Advantages
The capital costs for a diesel engine, generator and pump are acceptable in areas of high-powered pumping need, where no grid electricity is available.

Disadvantages
Frequent maintenance is required. Fuel cost may be high, and diesel fuel may be difficult to get. From time to time, a specialist mechanic is needed to service and repair the engine.

- **Solar cell systems**

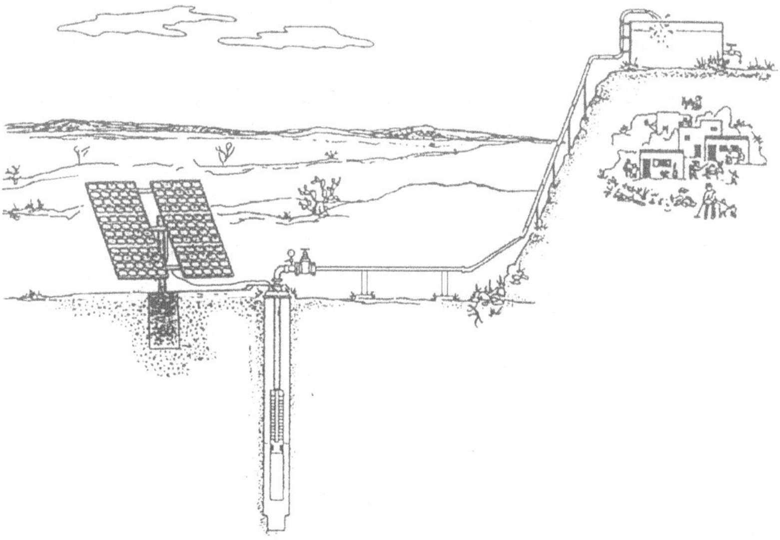

Source: SKAT (2000–2001)

Description
Photovoltaic or solar cells convert energy from light directly into electricity. They are made from special materials, such as silicon, germanium and selenium. A number of solar cells wired together under a protective glass plate are called a module. Modules can be connected in parallel or series, according to the required voltage and current. The electricity produced may go directly to an engine or be stored in batteries.

Operation
Solar cell systems work completely automatically. No operation is needed.

Maintenance
Periodic removal of dust from the glass module plates is about the only regular maintenance required. The external wires, the support structure of the array, the cover of the electronic components and, possible, a protective fence may need some occasional repairs.

Advantages
Solar cell systems are practicable in areas where sufficient sunlight is available, especially for small scale activities and where no grid electricity is available. The system requires little operation and maintenance.

Disadvantages
Initial investment is high. Solar cell systems are not feasible in areas with daily solar energy below 3 KWh/m^2.

- **Windmill**

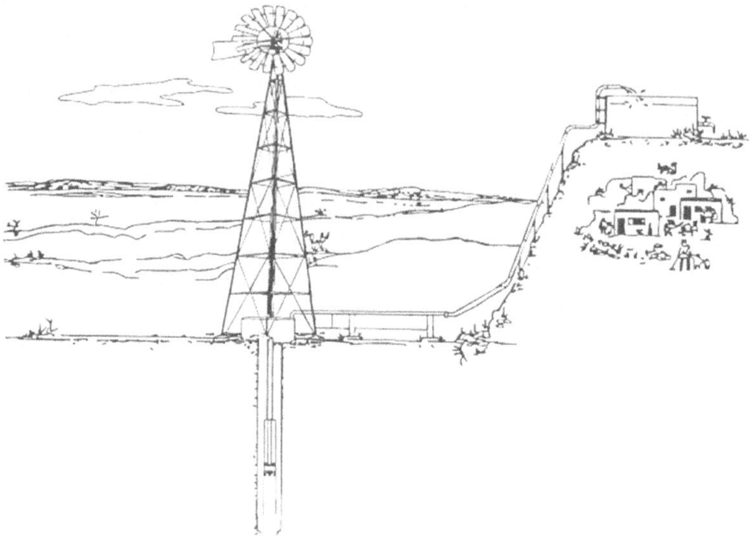

Source: SKAT (2000–2001)

Description
The most common models of windmill have a rotor fixed to a horizontal axis and are mounted on a tower of steel. Wind forces move the rotor and this movement drive a pump, usually of the piston type. Transmission can be direct or via a gearbox. A vane keeps the rotor facing the wind and positions it parallel to the wind in case of excessive wind speeds.

Operation
Some windmills require manual release of the furling mechanism after excessive wind. When no pumping is needed, the windmill may be temporarily furled out of the wind by hand. Windmill and pump functioning should be checked regularly to correct any abnormality.

Maintenance
Every month, the windmill and pump must be checked visually. Loose nuts and bolts have to be tightened, and moving parts may need greasing. Especially the bolts on pumping rods tend to loosen. Poor maintenance will lead to rapid bearing wear, and wind damage to the rotor and other parts.

Advantages
In windy areas where fuel is expensive or the supply is unreliable, windmill-activated pumps have proven to be a very competitive alternative to diesel-driven pump systems.

Disadvantages
When wind speeds are lower than 2 m/s, most windmills cannot pump. Badly trained caretakers may block the furling mechanism. Excessive winds may damage the windmill. Moving parts may wear fast because of lack of lubrication.

I.5 WATER TREATMENT TECHNOLOGIES

One of the most important objectives in any water-supply project is to select a source that is economically feasible and that has the best quality of water, so as to minimize dependence on treatment. The selection and design of treatment processes is based on the principles of keeping them as simple as possible, and guaranteeing their proper operation with minimal attention. The technology necessary for water treatment in rural communities is well established, and it is rare for innovative treatments to be more advantageous.

The risk of good quality water, either natural or treated, becoming contaminated in the household because of poor handling or storage must always be considered.

I.5.1 Treatment of groundwater

Groundwater is the preferred source of water for rural communities because it is generally of high quality and requires little treatment. A summary follows of the types of treatment processes most likely to be required for groundwater supplies.

- *No treatment*
  Gravity springs and deep well supplies may be quite safe without disinfection, if the source can be protected against contamination. Safety needs to be confirmed by microbiological and chemical testing of water quality.
- *Disinfection only*
  Groundwater, with disinfection as the only treatment, is the most common type of supply for small communities. Calcium hypochlorite or bleaching powder is the most appropriate disinfectant. On-site hypo-chlorination is feasible if units are manufactured locally and experience of their use has shown them to be reliable.
- *Iron removal*
  Many groundwater sources are high in iron and/or manganese. These elements pose no health problems, but their removal is often desirable because they discolour the water, as well as clothes and rice. The most common treatment is aeration, generally by spraying, followed possibly by sedimentation if the iron content is very high, and slow or rapid filtration. Manganese removal may often be improved by chlorination before filtration. If the source is protected, and the treatment facilities enclosed, disinfection may not be necessary.
- *Fluoride and arsenic removal*
  Long-term exposure to fluorides and arsenic in drinking-water is a serious health risk. Where fluorides need to be removed, the most suitable approaches are by ion exchange in activated aluminium or bone char filters, or by coagulation with aluminium. Defluoridation is costly and not always effective, so selection of an alternative source, even at greater cost, may be the most feasible solution.

  Arsenic toxicity has negative side-effects on health. Hence, provision of arsenic-free water sources (surface water, rainwater) is needed to mitigate the arsenic toxicity of people living in acute arsenic areas. Treatment of arsenic-contaminated water is not cheap, and it requires a major technological shift in water supply, such as co-precipitation and adsorption processes, lime treatment, ion exchange, membrane techniques and microbial processes.
- *Softening*
  Hard water creates difficulties for washing laundry. With the availability of modern detergents for washing, softening is not likely to be justified, given its high cost. Recent evidence on health benefits of hard water also makes softening less desirable from a public health perspective.

I.5.2 Treatment of surface water

A summary follows of the types of treatment processes most likely to be required for surface supplies.

- *Disinfection only*
  All surface waters must undergo at least disinfection. If the waters meet turbidity and colour standards without other treatment, disinfection alone will suffice. Where surface waters are highly coloured or turbid, resulting in a high chlorine demand, treatment as discussed below may be justified prior to disinfection. Where water is polluted with wastewaters, treatment beyond disinfection is certainly required.
- *Slow sand filtration*
  Where surface water exceeds turbidity standards, slow sand filtration is appropriate. Such filtration requires no pre-treatment, is simple in construction and operation, low in cost and is well suited to rural communities. Roughing filters may be useful to prepare waters with slight excess turbidity. Disinfection is required after filtration.
- *Rapid sand filtration*
- Most river supplies or other surface supplies that are high in turbidity, even if only during the wet season, require rapid sand treatment, along with pre-treatment that generally consists of chemical coagulation, flocculation and sedimentation. Slow sand filters clog too quickly.
- *Activated carbon treatment*
  Should it be necessary to draw water from a source of water having excessive taste and odour, or from a highly polluted source, activated carbon treatment may be useful. Carbon is added in a powdered formulation with the coagulant to absorb tastes and odours. It is not too expensive, because it needs to be used only during those periods when tastes and odours are troublesome.

I.5.3 Fact sheets

I.5.3.1 Household water treatment systems

Three household water treatment methods are addressed in these fact sheets: boiling, the use of a slow sand filter and chlorination. Other methods include the use of SODIS bottles (exposing drinking water in recycled PET bottles to six hours of UV radiation from sunlight) and the use of ceramic filters. Often, combinations of methods are used.

- **Boiling**

Source: Brikké et al. (1997)

Description
Boiling is an effective way to free water from micro-organisms. Usually it is recommended to let water boil for at least ten minutes at sea level and for a longer time at higher altitudes, where boiling temperatures are lower. High turbidity does not affect disinfection by boiling but if the water is to be filtered, this must be done before boiling. For household use, water is mostly boiled in a pot on a stove.

If it is not stored in the same pot in which it was boiled, the water should be poured into a clean storage container immediately after boiling in order to let the heat of this water kill bacteria that may be present in the storage container.

Operation
Boiling water on a kitchen stove is a task usually done by women. Water is put in a clean pot, which is placed on the lit stove. Sometimes herbs are added to the water. When the water starts to boil, the amount of fuel can be reduced. After boiling for some minutes, the water is allowed to cool down.

Maintenance
Maintenance of the stove and pots forms part of everyday maintenance of the kitchen. Depending on design, quality of materials and workmanship, and intensity of use, the stove has occasionally to be repaired or replaced.

Advantages
No extra investments apart from a container for the boiled water. Disinfection is organized within the household.

Disadvantages
Costs depend very much on fuel prices and stove efficiency. There is a risk of re-contamination of boiled water in conditions of poor hygiene. A change in the taste of water after boiling may affect its acceptability. Practised on a wide scale, boiling may contribute to carbon emissions.

- **Domestic slow sand filter**

Source: Brikké et al. (1997)

Description
In a household slow sand filter, water is purified through a combination of biological, physical and chemical proc sses that occur when water slowly passes downwards through a bed of sand. Fine particles are filtered out while the water is in the sand. Also, a population of micro-organism develops on top of the filter bed, and these microorganisms feed on bacteria, viruses and organic matter in the water, which helps to purify the water.

The system consists of a raw water supply, a filter tank and a clean water tank.

Operation
Operation of a slow sand filter is crucial to its effectiveness. The flow of water must be maintained constantly at 0.1 m per hour to provide the organisms in the filter with a stable flow of nutrients and oxygen, and give them time to purify the water. Flow is regulated by adjusting the floating weir. The raw water storage tank must never run dry.

Maintenance
After a few weeks to months of operation, depending on raw water quality, the flow rate in the filter gets too low and 1 or 2 cm of sand and organic material have to be scraped off the top of the filter, washed, dried in the sun and put aside. When the filter bed becomes too thin, the washed sand is restored. Every year the tank must be checked for corrosion.

Advantages
With good operation and maintenance, a household slow sand filter produces water virtually free from disease-causing organisms.

Disadvantages
In some regions, sand is expensive or difficult to get. As an alternative, material such as burnt rice husks could be used. Household slow sand filters require a substantial investment, and dedicated operation and maintenance, and as a result can be quite expensive to run.

- **Domestic chlorination**

Source: Brikké et al. (1997)

Description
Chlorination of water at household level can be applied as an emergency measure or as part of everyday life. When water quality cannot be trusted, a fixed amount of a concentrated chlorine

solution is added to a container with a fixed amount of clear water. The amount of chlorine needed depends mainly on the concentration of organic matter in the water and has to be determined for each situation.

Operation
A concentrated chlorine solution has to be made freshly at least every week but preferably more frequently. Chlorine-producing chemicals should be stored in a cool and dry place, and treated carefully. Contact with eyes or clothes should be avoided. The method to be used for disinfection with chlorine can easily be learned. Some education of people at the household level on this practice will be useful.

Maintenance
Apart from the cleaning and occasional replacement of containers and utensils, no maintenance is needed.

Advantages
All costs are recurrent costs and the cost of chlorine is relatively low. All tasks can be performed at the household level.

Disadvantages
Chlorination does not kill all pathogenic organisms, but it is generally very effective. Water containing suspended solids is not suitable for chlorination. Cost and unavailability of chlorine can be serious limitations.

I.5.3.2 Central water treatment systems

- **Pot chlorination in well**

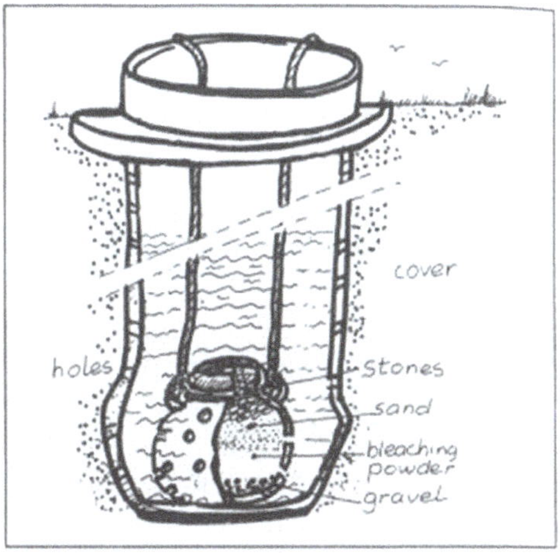

Source: Brikké et al. (1997)

Description
Pot chlorination is a method of continuous disinfection of the water in a dug well, spring or storage tank. A pot made of ceramics, plastic or other suitable material is filled with a mixture of two or three parts of sand and one of bleaching powder, sealed on the top with plastic and placed under

the water level. Sometimes the pot is placed inside another pot to slow down diffusion. The size of the pot and the number and size of the holes have to be adapted to the quantity of the water and the amounts used daily.

Operation
Every month, depending, among other factors, on water quality and quantity used, the pot is taken out of the water and the mixture of clean dry sand and bleaching powder is replaced. Then the pot is replaced in the well. The sand can be partly reused after washing and drying.

Maintenance
Apart from occasional replacement of the pot, plastic cover, rope and utensils, no maintenance is needed.

Advantages
Pot chlorination can be applied as an emergency measure or for longer periods. This method is suitable in areas where water sources need continuous disinfection and bleaching powder is available.

Disadvantages
When the water contains a lot of organic matter or suspended material, proper disinfection will require a lot of bleaching powder. Regular testing of residual chlorine in the water is needed. Cost and unavailability of bleaching powder can be serious limitations.

- **Storage and sedimentation**

Source: Brikké et al. (1997)

Description
The quality of raw water can be improved considerably by storage. A large storage reservoir makes it possible to avoid water intake when the quantity and quality of water at the source is temporarily low. A storage reservoir can be constructed in many ways: as an unlined pool at groundwater level; with a lining of loam, clay or concrete; or completely constructed in brick or concrete. Reservoirs for sedimentation commonly have two separate sections; while one is in use, the other can be cleaned.

Operation
Usually water will be let in every day or continuously, but when water quality becomes too low and sufficient storage exists, water intake may be interrupted temporarily.

Maintenance
Depending on silt content and reservoir depth, the reservoir will have to be flushed regularly in order to remove the deposited silt. All valves must be opened and closed at least once every two months to prevent them from getting stuck. Valves may occasionally need repair or replacement, and reservoir leaks will have to be fixed.

Advantages
Depending on the construction technology used, storage and sedimentation reservoirs can be used in areas where raw water contains many solids in suspension, or where quality or quantity at the source vary considerably.

Disadvantages
If the solids in the water do not settle fast enough, flocculation may be needed.

- **Up-flow roughing filter**

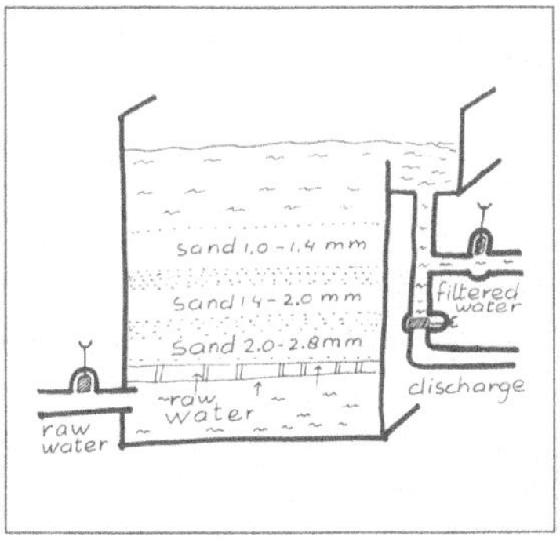

Source: Brikké et al. (1997)

Description
Roughing filters are often used as a pre-treatment technology, as they remove the suspended solids from the water that could rapidly clog a slow sand filter. They can remove a considerable amount of pathogens, iron and manganese. There are different types of roughing filters with different flow directions and with different types of filter medium.

For example, an up-flow filter box can be made of brickwork, concrete or Ferro-cement. Water flows in through a system of drains under the bottom of the filter box, usually a perforated PVC pipe, which also permits rapid extraction during cleaning (when the flow direction is reversed).

Operation
The filters are operated on a continuous basis. Operation consists of regulating the water flow and checking the turbidity of the effluent. If this turbidity gets too high or the filters get clogged, the filters should be cleaned.

Maintenance
Up-flow roughing filters are cleaned by backwashing and rapidly draining the water in the filter box. The surface layer of the filter has to be cleaned at least one a month; the inlet and outlet should also be cleaned. After several years of operation, hydraulic cleaning alone is no longer enough, and the different filter layers have to be removed and cleaned. All valves have to be opened and closed at least once every two months in order to prevent them from getting stuck.

Advantages
Filters are relatively cheap and easier to clean.

Disadvantages
Roughing filters only partly remove the solids and pathogens in water. Additional treatment is needed.

- **Slow sand filtration**

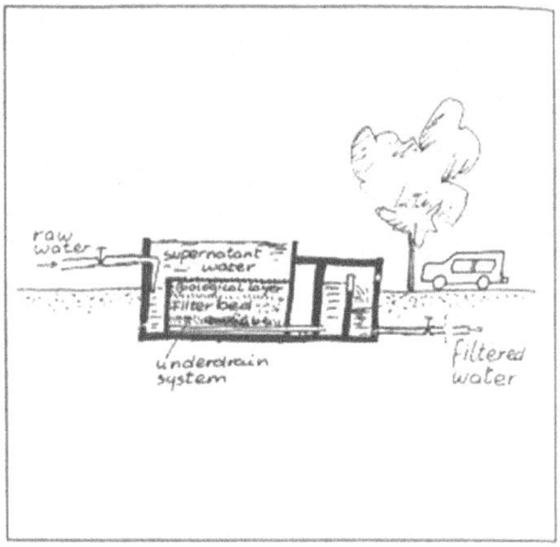

Source: Brikké et al. (1997)

Description
Slow sand filtration purifies water through a combination of biological, physical and chemical processes that occur when water slowly passes downwards through a bed of sand. Fine particles are filtered out, and in the sand and on top of the filter-bed, a population of microorganism develops that feeds on bacteria, viruses and organic matter in the water. The filter reservoirs have drains at the bottom, covered with gravel and the filter sand.

An inlet provides for the smooth entrance of raw water, and an outlet structure leads the clean water from the drains to the clean water mains.

Operation
Continuous operation of a slow sand filter is crucial for its effectiveness. The flow of water must be maintained in the range of 0.1 and 0.3 m per hour to provide the organisms in the filter with a stable flow of nutrients and oxygen, and give them time to purify the water. Flow rates may have to be adjusted accordingly, or the layer of supernatant water on the filter will get too high.

Maintenance
When flow velocities get too low, the filter has to be drained and the top layer of the sand scraped off, washed, dried in the sun and stored. Hygienic precautions must be taken every time someone enters a filter unit for maintenance or inspection. Valves must be opened and closed every two months to prevent them from getting stuck.

Advantages
With good operation and maintenance, a slow sand filter produces water virtually free from harmful organisms.

Disadvantages
In some regions, sand may be expensive or difficult to get. An alternative, such as burnt rice husks, may be used. Slow sand filters require a substantial investment, and dedicated operation and maintenance.

- **Chlorination in piped water supply systems**

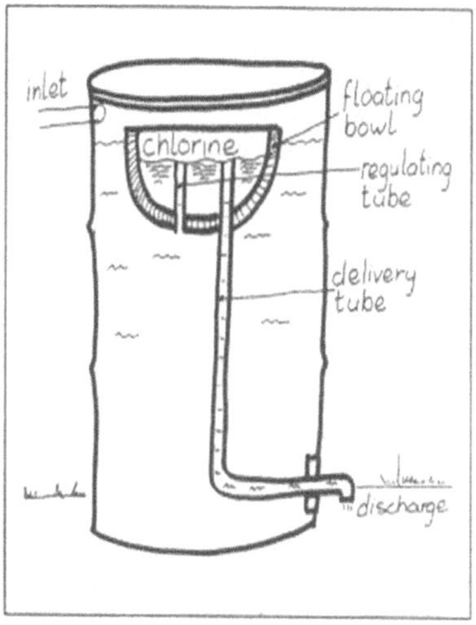

Source: Brikké et al. (1997)

Description
Chlorination is a chemical method of disinfecting water that kills nearly all pathogens and provides a barrier against re-infection. It can be applied as the last stage in a drinking-water treatment process or as the only measure when water quality is already reasonable good. The most commonly used low technology methods are batch chlorination and flow chlorination. For batch chlorination, a concentrated chlorine solution is added to the water in a reservoir, with the inlets and outlets closed. Flow chlorination continuously feeds small quantities of weakly concentrated chlorine solution to a flow of fresh water, often at the in-stream of a clear water reservoir.

Operation
The chlorine tank has to be refilled with a freshly prepared solution once or twice a week. The flow rate has to be checked, and adjusted if necessary. Operators must be very careful to avoid contact of chlorine compounds or solutions with eyes or clothes.

Maintenance
Chlorinators regularly have to be adjusted and cleaned of chlorine salts. Chlorine affects hoses, which then have to be replaced. If a steel chlorine tank is used, it must to be painted and checked for corrosion. Protective gloves and maintenance equipment will need to be replaced when they show signs of wear. The shelter of the tank in which the chlorine solution is kept needs maintenance.

Advantages
Recurrent costs for chlorine-producing chemicals are low, but this figure may vary substantially from one country to another. Costs for rubber gloves, hoses and other spare parts are generally low.

Disadvantages
Chlorination does not kill all pathogenic organisms, but it is generally very effective. When water contains a lot of organic matter or suspended material, pre-treatment will be needed. High cost and unavailability of chlorine compounds can be serious limitations.

I.6 WATER STORAGE SYSTEMS

I.6.1 Storage reservoirs

Water supply systems are well served by elevated service reservoirs to:

- maintain uniform pressure in the distribution system;
- store water for use during periods of peak demand;
- provide a reserve for supply during pump or power failures.

Including elevated storage in a system has the advantage of:

- permitting the pumps to operate at a low, steady rate, without the need for many on-and-off pump cycles, thereby lengthening the life of the engines;
- eliminating the need for pressure tanks to regulate pumping;
- making it feasible to use much smaller pumps, because they need only to provide the maximum daily flow – without elevated storage, the pumps need to provide for peak hourly flow;
- reducing the size of transmission lines, because they need only to provide maximum daily capacity and not peak hourly capacity;
- reducing the power capacity required for the pumps and the total power used, because the head losses in the system are lower at lower flow rates;
- reducing the required capacity of the clear well at a treatment plant, which is necessary to permit the plant to operate at a uniform rate;
- permitting the use of simple constant-feed chemical feed devices – an important consideration where chlorination is the only treatment;
- possibly providing contact time for the chlorine where there is no treatment plant;
- guaranteeing water supply during short periods of power failure, which are frequent in many rural communities;
- providing water for emergencies, such as during transmission line breaks, pump or source failures, and the like.

The storage capacity of elevated tanks is calculated on the basis of the volume required to make up for hourly variations in demand offset against the rate of pumping, plus a volume for emergencies. If the water supply has to cope with the demands of fire fighting, a suitable volume needs to be added to the designed storage capacity.

Care should, however, be taken to prevent tanks and reservoirs from becoming breeding places for malaria mosquitoes, especially in areas where malaria transmission is seasonal. Mosquito breeding in tanks and reservoirs may change the transmission pattern from seasonal to perennial. Storage tanks should therefore be covered, ventilated pipes should be screened with mosquito-proof mesh, and steps should be taken to avoid the creation of breeding sites downstream from the overflow. Special measures may need to be taken for raw water reservoirs (ponds and dams) to prevent their banks becoming overgrown with weeds, for instance by paving the banks at water level.

- *Location and elevation*
  Tank location and elevation affect pipe size and pump requirements. The best choice for service storage, if available, is a ground tank at an elevated location near the community to be served. If the land is flat, the storage tank must be elevated on a tower. It is often convenient to place the tank either at the site of the treatment plant or pumping station in order to facilitate operation and maintenance, or at a central location in the distribution network. If the area to be served stretches over a long distance, with the source at one end, the reservoir might be built at the other end. This allows the tank to be lower and the distribution pipes to be smaller, because water is fed to consumers from two directions. Such a location provides added reliability in the event of a break in the transmission main.
- *Storage technologies*
- Of the storage options reviewed below, the concrete-lined earthen reservoir is the best choice for raw water storage. In many cases, it can be used instead of an open concrete reservoir, and implications for operation and maintenance should be considered in the light of the water-lifting methods that make raw water storage necessary.

  Reservoir technologies not discussed here include fibreglass, break pressure, suspended crossing, valve boxes, frictional diffusers, and master water meters. Apparently, these technologies are less suitable in rural areas for low-income communities.

I.6.2 Fact sheets

- **Concrete-lined earthen reservoir**

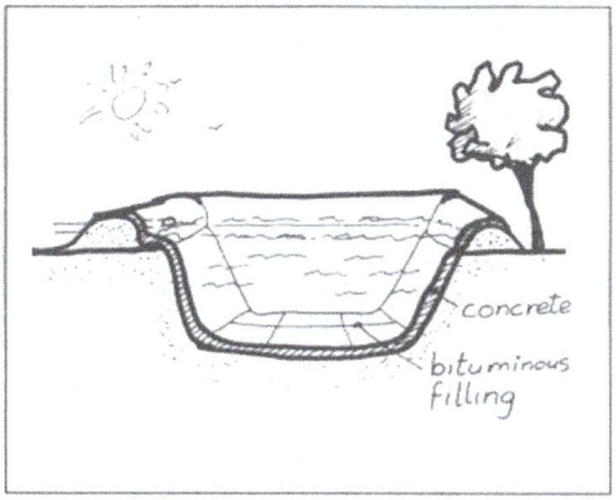

Source: Brikké et al. (1997)

Description
Lined earthen reservoirs can be built in natural depressions or constructed by a combination of excavation and building a dam around the reservoir. Inlets and outlets are installed during the earthworks. The reservoir is waterproofed by lining the inside with concrete, which is usually poured on site in big sections (several metres in width). A sealing of bituminous or other waterproof material is used to connect sections.

Operation
Operation of a reservoir is mainly a matter of opening and closing the inlet and outlet valves and sluices, according to the water needs of the system and the water quality at the inlet.

Maintenance
Valves and sluices have to be opened and closed once a month to prevent them from getting stuck. At least once a year, the reservoir has to be emptied of sediment, cleaned, and the lining inspected and disinfected with chlorine. Cracks or other damage to the lining have to be repaired.

Advantages
A lined earthen reservoir is the only technology suitable for raw water storage. Linings can also be made of clay, loam, plastic or asphalt.

Disadvantages
In regions with prolonged water shortages, the required size of a reservoir may be so big that the cost of building the reservoir may be beyond the financial resources of the community. Rapid filling with silt may be a problem, but this may be remedied by using raw water of an improved quality.

- **Reinforced concrete reservoir**

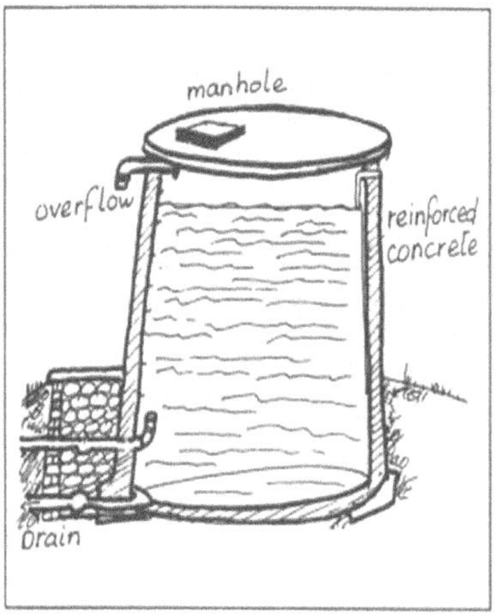

Source: Brikké et al. (1997)

Description
Reinforced concrete clear water reservoirs store clean water, to release it on demand. They are made of concrete reinforced with steel bars or steel mesh. Chemical additives are often mixed with the concrete to improve impermeability. Reinforced reservoirs are built on site, on a solid base of rock or undisturbed sand. To protect water from contamination, the reservoir is covered with a roof, which can be made of different materials, although reinforced concrete is usually used. At the top of the tank, an aeration pipe with a screen permits fresh air circulation in the tank, while keeping rodents and insects out. A manhole in the roof gives access for cleaning and repair.

Operation
Operation consists of opening and closing valves, and managing chlorine if provided. If the reservoir does not deliver directly to a tap, a caretaker usually operates the water supply.

Maintenance
A well-designed and well-built reservoir needs very little maintenance. The surroundings must be cleaned regularly. Every month, valves must be closed and opened to prevent them from getting

stuck, and screens must be checked and replaced as necessary. Once a year, or when contamination is suspected, the reservoir has to be drained, desalted, cleaned and disinfected with chlorine.

Advantages
Some construction technologies use bamboo or other local materials to reinforce concrete. Outlets are built a little above the floor of the reservoir, which has a slope pitched down towards one point where a washout pipe permits flushing.

Disadvantages
Reinforced concrete is expensive, and settlement of the soil under the structure may be great because of the heavy weight of concrete.

- **Elevated steel reservoir**

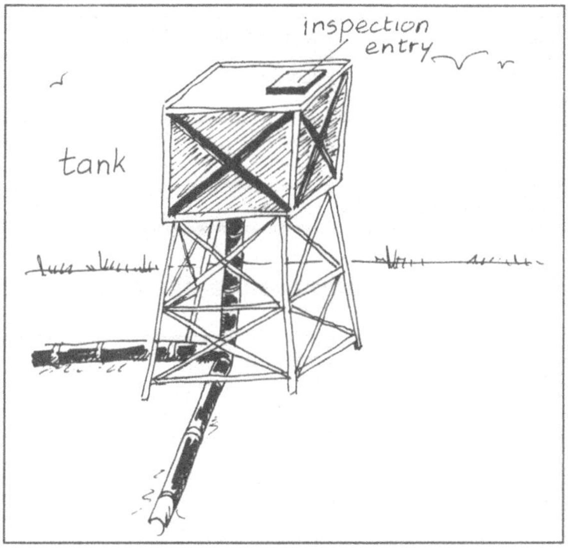

Source: Brikké et al. (1997)

Description
Elevated steel reservoirs store clean water in a steel tank on a high stand or tower. The high position of the tank provides the pressure needed to reach other points in the distribution system. For communal needs, elevated steel tanks are often constructed from galvanized steel elements bolted or welded together. Such tanks can be built faster and the transport cost of the material is generally lower than a concrete reservoir, when concrete materials are not locally produced. Several pipes are connected to the tank: inlet, outlet overflow and washout. An inspection entry is provided in the cover of the tank, and a screened vent hole or pipe maintains atmospheric pressure in the tank.

Operation
Operation consists of opening and closing valves, and managing chlorination if provided. It is usually carried out by a caretaker living nearby.

Maintenance

At least every six months, the inside of the reservoir has to be cleaned. Once a year, the paint on the tank needs to be touched up and the stand requires maintenance. Any leaks should be repaired immediately. Every time someone enters to the reservoir, it has to be completely disinfected using chlorine solution. Every two months, valves must be closed and opened to prevent them from getting stuck, and screens must be checked and replaced as necessary.

Advantages

The reservoir provides a stock of water for use during a dry period. The high position of the elevated steel reservoir provides the pressure needed to reach other points in the distribution system.

Disadvantages

Compared to concrete, Ferro-cement or even wood, steel reservoirs need relatively more maintenance.

- **Ferro-cement tank**

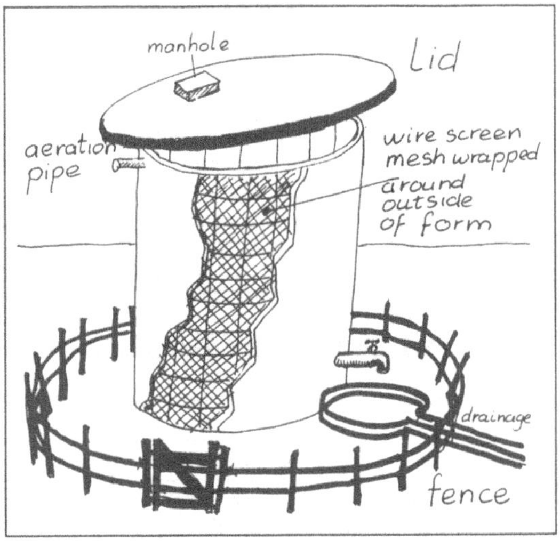

Source: Brikké et al. (1997)

Description

Ferro-cement water tanks are made of a structure of steel mesh and wire, covered on the inside and outside with a thin layer of cement-and-sand mortar. They can be used at household level or for complete communities, and provide a relatively inexpensive and easy-to-maintain type of storage. Compared to concrete reservoirs, Ferro-cement tanks are relatively light and flexible. To protect clear water from contamination, the tank is covered with a lid or a roof, which can be made of different materials, although Ferro-cement is usually used.

An aeration pipe with a screen permits fresh air circulation, while keeping rodents and insects out. Water flows into the reservoir through an inlet pipe that is above water level. This means that water cannot flow back into the pipe, and the sound of water entering the reservoir is easily audible.

To avoid bending forces in the wire meshes, most Ferro-cement tanks are cylinder-, globe- or egg-shaped.

Operation
Operation consists of periodic opening and closing of valves, and managing chlorination. It is usually done by a caretaker living nearby.

Maintenance
A well-designed and well-built tank needs very little maintenance. The surroundings, including the drain, must be kept clean. Every month, valves must be closed and opened to prevent them from getting stuck, and screens must be checked. Once a year, the reservoir has to be drained, desalted, cleaned and disinfected with chlorine.

Advantages
Compared to concrete reservoirs, Ferro-cement tanks are relatively light and flexible. Also, they are cheaper to build and require less qualified workers.

Disadvantages
Ferro-cement is less suitable for rectangular structures. In regions where Ferro-cement technology is unknown, it may take some time for people to accept such a different concept of construction. Generally, people do not trust the thin walls, and think that they require too much steel.

I.7 WATER DISTRIBUTION SYSTEMS

Ideally, drinking-water should be provided inside or near each household. If it is, this generally leads – over time – to an increase in the volume of water used. Even if only a single tap is installed, then this will have a notable impact on health and hygiene.

I.7.1 Distribution networks

Distribution systems are classified as branched networks or looped networks.

- **Branched networks**
  These networks are relatively easy to design and are less costly to construct in small systems. However, they are less reliable, as any break or maintenance work in the network results in cutting off the supply to part of the system. Also, the dead ends impair water quality. A network can be started as a branched system, and later on the ends of the branches can be "looped" to improve flow conditions and provide greater reliability.
- **Looped networks**
  These networks are common in densely populated areas. They provide greater reliability of services than branched networks, they eliminate dead ends, and they maintain stable flow conditions during periods of high local demand. Looped networks may be designed based on a main loop which feeds into secondary pipes or based on a branched network with several interconnections resulting in loops.

The cost of the distribution network depends mainly on the total length of pipes installed. Consequently, the design of the network should be carefully considered and the future development of the service area should be estimated.

Normally, pipes are situated in or along a main street, with branches into residential access roads. To make it easy to find pipes in order to carry out future repairs or checks, a standard layout system may be adopted, for example always laying the pipes on the same side of the road.

The depth at which the pipes are laid depends on the strength of the pipe materials, the traffic load and the depth of the frost line.

I.7.2 Fact sheets

- **Public stand-post**

Source: Brikké et al. (1997)

Description

At a public stand-post, people from several households can get water from one or more taps. Because stand-posts are used by many people and are often not well taken care of, their design and construction have to be sturdier than for domestic connections. The stand-post includes a service connection to the supplying water conduit, and a supporting column or wall with taps, to make it easy for people to fill water containers. The taps may be a globe or an automatic self-closing type. The column or wall may be made of wood, brickwork, dry stone masonry, concrete, or other material.

A solid stone or concrete slab or apron must be built under the tap, with a drainage system to lead spilled water away and prevent the formation of muddy pools. A fence may be needed to keep cattle away.

Operation

Water users clean and fill their containers at the tap. Bathing and washing of clothes is usually not permitted at the stand-post itself. The tap site has to be cleaned daily and the drain inspected.

Maintenance

The drain must be cleaned. Formation of pools must be prevented. Occasionally, the tubing may leak and need to be repaired or replaced.

Advantages

A lot of people can use the same tap, the maximum generally being 200 users per tap and depending on water yield.

Disadvantages

People have to be willing to organize communal use and maintenance. If they are willing to share and maintain the stand-post, then the only limitation is the cost. Special attention should be given to how water is handled after being collected at the stand-post in order to prevent subsequent contamination.

- **Domestic connection**

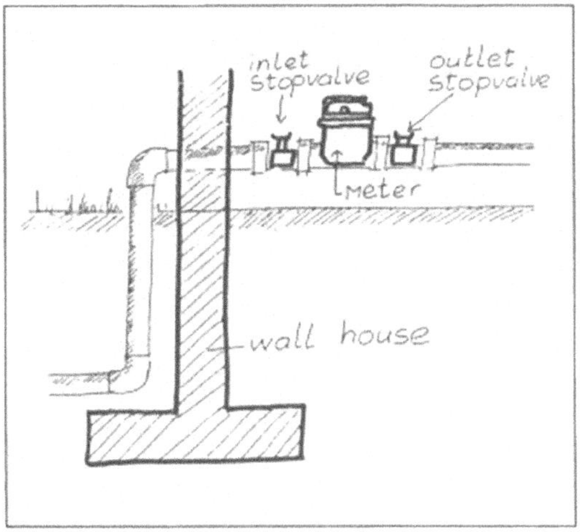

Source: Brikké et al. (1997)

Description

If enough water and funds are available, it may be possible to connect every house or yard to a piped water system. This is more convenient for water users, and generally increases water use and improves hygiene. A service pipe, usually made of polyethylene or PVC, leads from the distribution network to the house or the yard. The pipe leads either to a single tap on a post or to a system of pipes and taps in the house.

At the entry point to the premises, a gate valve and a water meter can be installed. Drainage facilities must also be provided.

Operation

Taps are used throughout the day. They should not be left open or leak. Formation of mud and pools by dripping taps must be avoided. The tap and site have to be cleaned daily and the drain inspected.

Maintenance

Once in a while, a rubber washer or other part of the tap may have to be replaced. Any structure on the tap site or the drainage system may need to be repaired. Occasionally, the tubing may leak and need to be repaired or replaced.

Advantages

A domestic connection is the most convenient water source for a household.

Disadvantages

Initial costs for household connections are higher than for shared stand-posts. Also, a network providing a higher number of connections, which is needed to provide domestic connections, is more complicated to maintain. If water is scarce, it may be difficult to guarantee water for all.

- **Small-flow water meter**

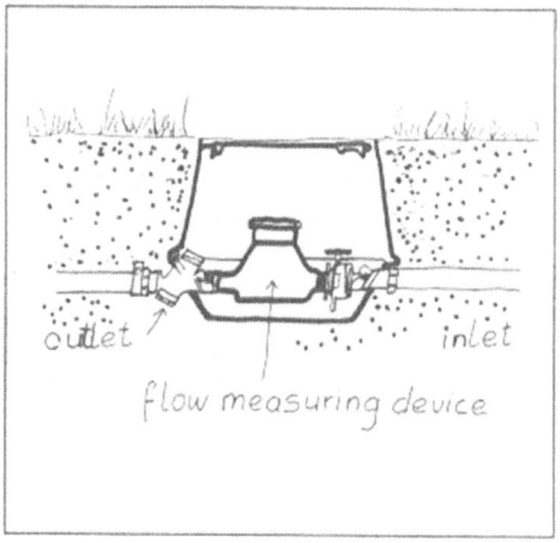

Source: Brikké et al. (1997)

Description
Water meters on public stand-posts or domestic connections provide a means of charging fees for water according to the volume of water delivered and thus of regulating water use via a price mechanism. A water meter consists of a flow-measuring device, a counter and a protective housing with an inlet and an outlet. A shut-off valve is always installed at the up-flow side of the meter to permit servicing.

Operation
Every month, the water meter is read by an appointed person, who writes down the new meter count in a book. The difference between two readings of the same meter is the amount of water used, and the consumer will be billed according to this quantity.

Maintenance
If the water is free of silt, a good quality water meter needs very little maintenance, but any repairs have to be done by a specialized workshop. When the strainer at the inlet gets clogged, cleaning can be done locally. It is advisable to clean the strainer at least one a year, depending on meter and water quality. If a water meter is not working, it has to be replaced or recalibrated. An old meter may be checked by a representative of the water committee against another meter known to be accurate.

Advantages
Households pay for their water consumption.

Disadvantages
Initial costs are high. The addition of water meters increases operational costs. Water meters are not suited for water with a high silt load.

Annex II

Project questionnaires

II.1 PROJECT SCENARIO QUESTIONNAIRE

Date: ____________________

1. Country: ____________________
2. Organization responsible for project implementation:

   ____
3. Name of the project (please indicate whether the project is planned, under construction or already implemented):

   ____
4. Design lifetime of the project: ____________________ years
5. Town or village where the water supply facility is (or will be) located:

   ____
6. Scenario of the size of the population supplied with drinking-water by the extended or new water supply facility:
   - Initial value: ____________________ inhabitants
   - Design value: ____________________ inhabitants
   - Trend duration: ____________________ years
   - Time trend shape:

     ☐ Constant

     ☐ S-shaped (alpha = beta = 2)

     ☐ Linear-shaped (alpha = beta = 1)

     ☐ J-shaped (alpha = 2, beta = 1)

     ☐ Rotated J-shaped (alpha = 1, beta = 2)

☐ Free Beta-shaped with: alpha = __________ and beta = __________
☐ Free trend scenario (specify): __________

7. Scenario of the average size of the household supplied with drinking-water by the extended or new water supply facility:
   - Initial value: __________ inhabitants per household
   - Design value: __________ inhabitants per household
   - Trend duration: __________ years round
   - Time trend shape:
     ☐ Constant
     ☐ S-shaped (alpha = beta = 2)
     ☐ Linear-shaped (alpha = beta = 1)
     ☐ J-shaped (alpha = 2, beta = 1)
     ☐ Rotated J-shaped (alpha = 1, beta = 2)
     ☐ Free Beta-shaped with: alpha = __________ and beta = __________
     ☐ Free trend scenario (specify): __________
     ☐ Not available (N/A)

8. Scenario of the average quantity of the per capita drinking-water supplied by the extended or new water supply facility:
   - Initial value: __________ litres per day per inhabitant
   - Design value: __________ litres per day per inhabitant
   - Trend duration: __________ years round
   - Time trend shape:
     ☐ Constant
     ☐ S-shaped (alpha = beta = 2)
     ☐ Linear-shaped (alpha = beta = 1)
     ☐ J-shaped (alpha = 2, beta = 1)
     ☐ Rotated J-shaped (alpha = 1, beta = 2)
     ☐ Free Beta-shaped with: alpha = __________ and beta = __________
     ☐ Free trend scenario (specify): __________
     ☐ Not available (N/A)

9. Improved drinking-water technologies being considered for the project:
   ☐ Rainwater collection
   ☐ Protected spring
   ☐ Protected dug well
   ☐ Tube well or borehole
   ☐ Public tap or standpipe
   ☐ Piped water into dwelling, plot or yard
   ☐ Others (specify): __________

10. Type of water supply source (or sources) available:
    ☐ Surface water
    ☐ Groundwater

☐ Rainwater
☐ Others (specify): ____________________
Name and signature of the person who filled in the questionnaire:
____

II.2 TECHNICAL QUESTIONNAIRE

Date: ____________________

GENERAL INFORMATION

1. Country: ____________________
2. Region: ____________________
3. Location of project (town or village): ____________________
4. Source of data: ____________________
5. Number of inhabitants: ____________________
6. Number of dwellings/houses: ____________________
7. Number of families: ____________________
8. Number of persons per family: ____________________
9. Density (inhabitants per km^2): ____________________

COMMUNITY CHARACTERISTICS

10. Description of the community: ____________________
____
____

11. Type of dwellings/houses: ____________________
____
____

12. Schools, postal office, public services: ____________________
____
____

13. Equipment of the community: ____________________
____

14. Distance to the nearest community: __________ km

15. Quality of the communications (highway, roads, etc.):
☐ Very good
☐ Good
☐ Average
☐ Poor
☐ Very poor

16. Type of transport (indicate all options):
    - ☐ Automobile
    - ☐ Bus
    - ☐ Train
    - ☐ Others (specify): ______________________________

17. Current water supply system:
    - ☐ Surface water (river, dam, lake, pond, stream, canal, irrigation channels)
    - ☐ Tanker-truck
    - ☐ Bottled water
    - ☐ Cart with small tank/drum
    - ☐ Unprotected spring
    - ☐ Unprotected dug well
    - ☐ Rainwater collection
    - ☐ Protected spring
    - ☐ Protected dug well
    - ☐ Tube well or borehole
    - ☐ Public tap or standpipe
    - ☐ Piped water into dwelling, plot or yard
    - ☐ Others (specify): ______________________________

18. Type of source for current water supply:
    - ☐ Surface water
    - ☐ Groundwater
    - ☐ Rainwater
    - ☐ Others (specify): ______________________________

19. Current sanitation system:
    - ☐ Open defecation
    - ☐ Hanging latrine
    - ☐ Bucket latrine
    - ☐ Pit latrine without slab or platform
    - ☐ Flush or pour–flush to elsewhere
    - ☐ Composting toilet
    - ☐ Pit latrine with slab
    - ☐ Ventilated improved pit latrine
    - ☐ Flush or pour–flush to:
    - ☐ Piped sewer system
        - ☐ Septic tank
        - ☐ Pit latrine
        - ☐ Others (specify): ______________________________

20. Company responsible for power (energy): ______________________________
    ____

21. Power installed: ____________ KWh

22. Power provided (on average, per 24-hour period):
    From ______ : ______ to ______ : ______

23. Map of the region (specify):

HYDROLOGY AND CLIMATE

24. Geographical location:
    - ☐ Coastal
    - ☐ Continental
    - ☐ Forest
    - ☐ Mountain
    - ☐ Others (specify) ______________________________

25. Average temperatures (in degrees Celsius):

☐ Winter	____________	☐ Dry season	____________
☐ Spring	____________	☐ Intermediate season	____________
☐ Summer	____________	☐ Rainy season	____________
☐ Fall	____________		

Minimum and maximum temperatures (in degrees Celsius):

☐ Winter	________/________	☐ Dry season	________/________
☐ Spring	________/________	☐ Intermediate season	________/________
☐ Summer	________/________	☐ Rainy season	________/________
☐ Fall	________/________		

26. Average rainfall (in millimetres per year):

☐ Winter	____________	☐ Dry season	____________
☐ Spring	____________	☐ Intermediate season	____________
☐ Summer	____________	☐ Rainy season	____________
☐ Fall	____________		

27. Rainy months:
    - ☐ January
    - ☐ February
    - ☐ March
    - ☐ April
    - ☐ May
    - ☐ June
    - ☐ July
    - ☐ August
    - ☐ September
    - ☐ October
    - ☐ November
    - ☐ December

28. Dry months:
    - ☐ January
    - ☐ February
    - ☐ March
    - ☐ April
    - ☐ May
    - ☐ June
    - ☐ July
    - ☐ August
    - ☐ September
    - ☐ October
    - ☐ November
    - ☐ December

Name and signature of the person who filled in the questionnaire:

____

II.3 SOCIOECONOMIC QUESTIONNAIRE

Date: ________________

GENERAL INFORMATION

1. Country: ____________________
2. Region: ____________________
3. Location of project (town or village): ____________________
4. Source of data: ____________________
5. Number of inhabitants: ____________________
6. Number of dwellings/houses: ____________________
7. Number of families: ____________________

8. Number of persons per family: ______________________________
9. Density (inhabitants per km^2):______________________________

COMMUNITY CHARACTERISTICS

10. Type of community:
    - ☐ Urban
    - ☐ Suburban
    - ☐ Rural

11. Proportion of men and women:
    Men ____________%
    Women ____________%

12. Demographic structure of the population:
    <5 years ____________%
    5–12 years ____________%
    13–18 years ____________%
    19–45 years ____________%
    46–65 years ____________%
    >65 years ____________%

13. Proportion of population according to highest educational level:
    - ☐ Complete university education ____________%
    - ☐ Incomplete university education ____________%
    - ☐ Complete secondary education ____________%
    - ☐ Incomplete secondary education ____________%
    - ☐ Complete primary education ____________%
    - ☐ Incomplete primary education ____________%
    - ☐ No education ____________%

14. Level of income according to population percentile:
    - ☐ 1% to 20% ____________ US$ or local currency
    - ☐ 20% to 39% ____________ US$ or local currency
    - ☐ 40% to 59% ____________ US$ or local currency
    - ☐ 60% to 79% ____________ US$ or local currency
    - ☐ 80% to 100% ____________ US$ or local currency
    - ☐ Subsistence income

15. Type and level of employment:
    - ☐ Formal ____________% ☐ Seasonal ____________%
    - ☐ Informal ____________% ☐ Full-time ____________%

16. Proportion of employment by production sector:
    - ☐ Agriculture, cattle and livestock ____________%
    - ☐ Commerce and industry ____________%

- ☐ Transport and telecommunication ____________%
- ☐ Mining ____________%
- ☐ Forestry ____________%
- ☐ Tourism ____________%
- ☐ Housing and construction ____________%
- ☐ Services ____________%
- ☐ Others (specify): ____________ ____________%

SOCIAL ORGANIZATIONS/ACTIVITIES

17. Local organizations:

	Yes	No
– Neighbourhood associations'	____________	____________
– Farmers associations'	____________	____________
– Parents and family centres	____________	____________
– Sports clubs	____________	____________
– Gender groups	____________	____________

18. Names of social organizations:

____

____

19. Attitude of the community to the sanitation:

____

____

20. Possible contributions of the community to the water supply project:
    - ☐ Financial resources ____________% or ____________ local currency or US$
    - ☐ Materials ____________% or ____________ local currency or US$
    - ☐ Labour ____________% or ____________ hours
    - ☐ Equipment ____________% or ____________ local currency or US$
    - ☐ Others (specify): ____

      ____________% or ____________ hours or local currency or US$

Name and signature of the person who filled in the questionnaire:

____

Annex III

Inputs of improved water supply technologies

III.1 PIPED WATER INTO DWELLING, PLOT OR YARD

INVESTMENT COSTS			
ACTIVITY	**ITEM**	**SUB-ITEM**	**INPUT**
MATERIALS			
Water collection	Well excavation and drilling	Hand dug well	Various tools (shovel, pick, spud bar and others)
		Driven well	Sledgehammer, drive cap, metal well point
		Jetted well	Tripod and pulley, hollow drill (rod in sections), pump, plastic pipe
		Bored well	Tripod, handle, coupling, auger iron, plastic or asbestos pipe
	Well lining	Hand dug well	Stone, brick or concrete (headwall), concrete rings, masonry with bricks or concrete blocks, clay, backfill (lining), gravel or prefabricated caisson intake
		Drilled well	Concrete, PVC pipe, gravel
Water conveyance	Pumps	Diesel pump	Superstructure (wood, slate, others), cables, pipes, and so on.
		Electric pump	Superstructure (wood, slate, others), cables, pipes, and so on.
		Solar pump	Cables, pipes, and so on.
		Wind pump	Cables, pipes, and so on.
	Transport	Pipes	Concrete, PVC or other
Water treatment	Central treatment	Storage and sedimentation	Unlined pool or dirt with a lining of plastic (loam, clay, concrete, brick), chemicals

(*Continued*)

INVESTMENT COSTS			
ACTIVITY	**ITEM**	**SUB-ITEM**	**INPUT**
		Up-flow roughing filters	Sand, gravel, coconut husk fibre, brickwork, concrete or Ferro-cement, PVC pipe, valve, plastic wire mesh
		Slow sand filtration	Sand, gravel, chemicals, valve, concrete, bricks, Ferro-cement, metal tube
		Chlorination in piped supply systems	Chlorine solution, valve, PVC tube
	Household treatment	Slow sand filter	Sand and gravel
		Chlorination	Chlorine solutions
Water storage	Storage tank	Concrete-lined earthen reservoir	Concrete, inlet and outlets, bituminous filling (or other waterproof materials) or Ferro-cement
		Reinforced concrete reservoir	Drain, concrete reinforced with steel bars or steel mesh, a roof of different materials (e.g. concrete reinforced), aeration pipe with a screen, a manhole, an inlet and outlets
		Elevated reservoir (tower)	For family use, the tank can be made of an oil-drum and the tower of bamboo or wood
			For communal use, the tank can be made with galvanized steel and the tower of steel, wood, or reinforced concrete. Several pipes and an inspection entry are needed
		Ferro-cement tank	Aeration pipe, manhole, lid, wire screen mesh wrapped around outside of form, drainage and fence
Water distribution	Private connection	Connection	Pipe (PVC or other), tap, cement, sand
		Trench	Various tools (shovel, pick, spud bar and other)
EQUIPMENT			
Water collection	Well excavation	Well drilling	Hydraulic rotary rigs or percussion rigs
Water conveyance	Pumps	Diesel pump	Motor, engine
		Electric pump	Motor, engine, control panel
		Solar pump	Photovoltaic or solar cells
		Wind pump	Rotor fixed to a horizontal axis and mounted on a tower of steel. A gear box is needed
		Axial flow pump	Often a boat propeller is used to improvise an axial flow pump. It turns in a cylindrical casing where fixed guide blades may be installed

INVESTMENT COSTS			
ACTIVITY	**ITEM**	**SUB-ITEM**	**INPUT**
		Centrifugal pump	The essential components are the fast rotating impeller and the casing
		Submersible pump	Pump and motor
		Hydraulic ram pump	Drive pipe, delivery valve, air chamber
Water treatment	Central treatment	Chlorination in piped supply systems	Reservoir tank
	Household treatment	Slow sand filter	Raw water tank, filter tank and clean water tank
Water distribution	Private connection	Trench	Bulldozer (rented - including operator). Compactor machine (rented - including operator)
		Supporting	Water meter
LABOUR			
Water collection	Well excavation and drilling	Hand dug well	Unskilled worker and skilled supervisor
		Driven well	Unskilled worker and skilled supervisor
		Jetted well	Unskilled worker and skilled supervisor
		Bored well	Unskilled worker and skilled supervisor
		Drilled well	Skilled worker
	Well lining	Hand dug well	Skilled worker or semi-skilled
		Drilled well lining	Skilled worker or semi-skilled
Water conveyance	Power systems	Diesel, electric, solar and wind	Skilled worker
	Power pump	Axial flow pump	Plumber
		Centrifugal pump	Plumber
		Submersible pump	Plumber
		Hydraulic ram pump	Plumber
	Transport	Pipes	Skilled worker
Water treatment	Central treatment	Storage and sedimentation	Plumber
		Up-flow roughing filters	Plumber
		Slow sand filtration	Plumber
		Chlorination in piped supply systems	Plumber
	Household treatment	Slow sand filter	Member of the family

(*Continued*)

INVESTMENT COSTS			
ACTIVITY	**ITEM**	**SUB-ITEM**	**INPUT**
		Chlorination	Member of the family
Water storage	Storing tank	Concrete-lined earthen reservoir	Skilled worker
		Reinforced concrete reservoir	Skilled worker
		Elevated reservoir (tower)	Skilled worker
		Ferro-cement tank	Skilled worker
Water distribution	Private connection	Connection	Plumber
		Trench	Semi-skilled worker
OPERATION AND MAINTENANCE COSTS			
ACTIVITY	**ITEM**	**SUB-ITEM**	**INPUT**
MATERIALS			
Water collection	Well disinfection	Hand dug well	Chlorine (or other chemical product)
	Well lining	Hand dug well	Replacement of the apron, headwall, drain
			Cleaning of the apron, headwall, drain
			Extension of the well
		Drilled well lining	Replacement of the apron, headwall, drain
			Cleaning of the apron, headwall, drain
			Rehabilitation of the well (various technologies such as air forced and water pumping, brushing and treatment with chemicals)
Water conveyance	Power system	Diesel	Cleaning of the structure
			Checking and replacement of power system pieces
			Fuel, oil
		Electric	Cleaning of the structure
			Checking and replacement of power system pieces
		Solar	Cleaning of the structure
			Checking and replacement of power system pieces
		Wind	Cleaning of the structure
			Checking and replacement of power system pieces
	Transport	Pipes	Replacement of pipes

OPERATION AND MAINTENANCE COSTS			
ACTIVITY	**ITEM**	**SUB-ITEM**	**INPUT**
Water storage	Storage tank	Concrete-lined earthen reservoir	Replacement of pieces, cleaning, checking the valves, screens and lining
			Draining and disinfecting the reservoir
		Reinforced concrete reservoir	Replacement of pieces, cleaning, checking the valves and screens
			Draining and disinfecting the reservoir
		Elevated reservoir (tower)	Replacement of pieces, cleaning, checking the valves and screens
			Draining and disinfecting the reservoir
			Painting the tank and tower
		Ferro-cement tank	Replacement of pieces, cleaning, checking the valves and screens
			Draining and disinfecting the reservoir
Water treatment	Central treatment	Storage and sedimentation	Regulation of inlet
			Flushing of deposited silt
			Repair or replacement of valve
		Up-flow roughing filters	Regulation of flow
			Cleaning of site
			Greasing of valve
			Repair or replacement of valve
		Slow sand filtration	Checking of flow
			Cleaning of site
			Scraping off sand, washing, drying and storage
			Re-sanding of filter
			Repair or replacement of valve and metal tube
			Disinfection of filter outlet
		Chlorination in piped systems	Refilling of chlorine tank
			Cleaning and replacement of chlorinator
			Checking and adjustment of doses
	Household treatment	Slow sand filter	Cleaning or replacement
		Chlorination	Cleaning
Water distribution	Private connection	Connection	Replacement of pieces and pipe
		Trench	Repair

(*Continued*)

OPERATION AND MAINTENANCE COSTS			
ACTIVITY	**ITEM**	**SUB-ITEM**	**INPUT**
EQUIPMENT			
Water conveyance	Power pump	Axial flow pump	Repair or replacement
		Centrifugal pump	Repair or replacement
		Submersible pump	Repair or replacement
		Hydraulic ram pump	Repair or replacement
Water treatment	Household treatment	Slow sand filter	Replacement of pieces
		Chlorination	Replacement of pieces
	Central treatment	Storage and sedimentation	Repair or replacement of pieces
		Up-flow roughing filters	Repair or replacement of pieces
		Slow sand filtration	Repair or replacement of pieces
		Chlorine in piped systems	Repair or replacement of pieces
			Refilling of chlorine tank
Water distribution	Private connection	Connection	Replacement of the water meter
LABOUR			
Water collection	Well disinfection	Hand dug well	Unskilled worker
	Well lining	Hand dug well	Repairs (semi-skilled worker)
			Extension (skilled worker)
			Cleaning (unskilled worker)
		Drilled well lining	Repairs (semi-skilled worker)
			Rehabilitation (skilled worker)
			Cleaning (unskilled worker)
Water conveyance	Power system	Diesel	Skilled worker
		Electric	Skilled worker
		Solar	Skilled worker
		Wind	Skilled worker
	Power pump	Axial flow pump	Plumber
		Centrifugal pump	Plumber
		Submersible pump	Plumber
		Diesel engine	Plumber

OPERATION AND MAINTENANCE COSTS			
ACTIVITY	**ITEM**	**SUB-ITEM**	**INPUT**
		Hydraulic ram pump	Plumber
	Transport	Pipes	Skilled worker
Water storage	Storage tank	Concrete-lined earthen reservoir	Semi-skilled worker (caretaker-inspection) and plumber
		Reinforced concrete reservoir	Semi-skilled worker (caretaker-inspection) and plumber
		Elevated reservoir (tower)	Semi-skilled worker (caretaker-inspection) and plumber
		Ferro-cement tank	Semi-skilled worker (caretaker-inspection) and plumber
Water treatment	Household treatment	Heating	Family member
		Slow sand filter	Family member
		Chlorination	Family member
	Central treatment	Storage and sedimentation	Semi-skilled worker (caretaker-inspection) and plumber
		Up-flow roughing filters	Semi-skilled worker (caretaker-inspection) and plumber
		Slow sand filtration	Semi-skilled worker (caretaker-inspection) and plumber
		Chlorination in piped systems	Semi-skilled worker (caretaker-inspection) and plumber
Water distribution	Private connection	Connection	Plumber
		Supporting	Skilled worker
		Trench	Semi-skilled worker

III.2 PUBLIC TAP OR STANDPIPE

INVESTMENT COSTS			
ACTIVITY	**ITEM**	**SUB-ITEM**	**INPUT**
Materials			
Water collection	Well excavation and drilling	Hand dug well	Various tools (shovel, pick, spud bar and others)
		Driven well	Sledgehammer, drive cap, metal well point
		Jetted well	Tripod and pulley, hollow drill (rod in section), pump, plastic pipe

(*Continued*)

INVESTMENT COSTS			
ACTIVITY	**ITEM**	**SUB-ITEM**	**INPUT**
		Bored well	Tripod, handle, coupling, auger iron, plastic or asbestos pipe
	Well lining	Dug well	Stone, brick or concrete (headwall). Concrete rings, masonry with bricks or concrete blocks, clay, backfill (lining), gravel or prefabricated caisson intake
		Drilled well lining	Concrete, PVC pipe, gravel
Water conveyance	Pumps	Diesel pump	Superstructure (wood, slate, others), cables, pipes, etc.
		Electric pump	Superstructure (wood, slate, others), cables, pipes, etc.
		Solar pump	Cables, pipes, etc.
		Wind pump	Cables, pipes, etc.
	Transport	Pipes	Concrete, PVC or other
Water storage	Storage tank	Concrete-lined earthen reservoir	Concrete, inlet and outlets, bituminous filling (or other waterproof materials) or Ferro-cement
		Reinforced concrete reservoir	Drain, concrete reinforced with steel bars or steel mesh, a roof of different materials (e.g. concrete reinforced), aeration pipe with a screen, a manhole, an inlet and outlets
		Elevated reservoir (tower)	For family use, the tank can be made of an oil-drum and the tower of bamboo or wood
			For communal use, the tank can be made with galvanized steel and the tower of steel, wood, or reinforced concrete. Several pipes and an inspection entry are needed
		Ferro-cement tank	Aeration pipe, manhole, lid, wire screen mesh wrapped around the steel structure with a layer of cement, drainage and fence
Water distribution	Public standpipe	Connection	Pipe (PVC or other), tap, cement, sand
		Supporting structure	Column in wall (wood, brickwork, dry stone masonry concrete), one (or more) globe or self-closing tap, regulating valve, a solid stone or concrete slab or apron, and a fence
		Spillage collector	Concrete slab, drain, soak-away (pit filled with stones and covered with a layer of soil)
		Trench	Various tools (shovel, pick, spud bar and other)
Water treatment	Household treatment	Slow sand filter	Sand and gravel
		Chlorination	Chlorine solutions

INVESTMENT COSTS			
ACTIVITY	**ITEM**	**SUB-ITEM**	**INPUT**
	Central treatment	Storage and sedimentation	Unlined pool or dirt with a lining of plastic (loam, clay, concrete, brick), chemicals
		Up-flow roughing filters	Sand, gravel, coconut husk fibre, brickwork, concrete or Ferro-cement, PVC pipe, valve, plastic wire mesh
		Slow sand filtration	Sand, gravel, chemicals, valve, concrete, bricks, Ferro-cement, metal tube
		Chlorination in piped supply systems	Chlorine solution, valve, PVC tube
Equipment			
Water collection	Well excavation	Well drilling	Hydraulic rotary rigs or percussion rigs
Water conveyance	Power system	Diesel power	Engine
		Electric power	Engine, control panel
		Solar power	Photovoltaic or solar cells
		Wind power	Rotor fixed to a horizontal axis and mounted on a tower of steel. A gearbox is needed
	Power pump	Axial flow pump	Often a boat propeller is used to improvise an axial flow pump. It turns in a cylindrical casing where fixed guide blades may be installed
		Centrifugal pump	The essential components are the fast rotating impeller and the casing
		Submersible pump	Pump and motor
		Hydraulic ram pump	Drive pipe, delivery valve, air chamber
	Manual pump	Suction plunger hand pump	Pump
		Direct action hand pump	Pump
		Deep well piston pump	Pump
		Deep well diaphragm pump	Pump
Water distribution	Public standpipe	Trench	Bulldozer (rented - including operator)
			Compactor machine (rented - including operator)
		Supporting	Water meter

(*Continued*)

INVESTMENT COSTS			
ACTIVITY	**ITEM**	**SUB-ITEM**	**INPUT**
Water treatment	Household treatment	Slow sand filter	Water raw tank, filter tank and clean water tank
	Central treatment	Chlorination in piped supply systems	Reservoir tank
Labour			
Water collection	Well excavation and drilling	Hand dug well	Unskilled worker and skilled supervisor
		Driven well	Unskilled worker and skilled supervisor
		Jetted well	Unskilled worker and skilled supervisor
		Bored well	Unskilled worker and skilled supervisor
		Drilled well	Skilled worker
	Well lining	Dug well	Skilled worker or semi-skilled
		Drilled well lining	Skilled worker or semi-skilled
Water conveyance	Power system	Diesel	Skilled worker
		Electric	Skilled worker
		Solar	Skilled worker
		Wind	Skilled worker
	Power pump	Axial flow pump	Plumber
		Centrifugal pump	Plumber
		Submersible pump	Plumber
		Hydraulic ram pump	Plumber
	Transport	Pipes	Skilled worker
Water storage	Storage tank	Concrete-lined earthen reservoir	Skilled worker
		Reinforced concrete reservoir	Skilled worker
		Elevated reservoir (tower)	Skilled worker
		Ferro-cement tank	Skilled worker
Water distribution	Public standpipe	Connection	Plumber
		Trench	Unskilled worker
		Supporting	Unskilled worker and plumber
		Collecting spillage	Skilled worker

INVESTMENT COSTS			
ACTIVITY	**ITEM**	**SUB-ITEM**	**INPUT**
		Wood fence	Unskilled worker
		Cement fence	Unskilled worker
Water treatment	Household treatment	Slow sand filter	Family member
		Chlorination	Family member
	Central treatment	Storage and sedimentation	Plumber
		Up-flow roughing filters	Plumber
		Slow sand filtration	Plumber
		Chlorination in piped supply systems	Plumber

OPERATION AND MAINTENANCE COSTS			
ACTIVITY	**ITEM**	**SUB-ITEM**	**INPUT**
Materials			
Water collection	Well disinfection	Dug well	Chlorine (or other chemical product)
	Well lining	Dug well	Repairing the apron, headwall, drain
			Cleaning the apron, headwall, drain
			Extension of the well
		Drilled well lining	Repairing the apron, headwall, drain
			Cleaning the apron, headwall, drain
			Rehabilitation of the well (various technologies such as air-forced and water pumping, brushing and treatment with chemicals)
Water conveyance	Power system	Diesel	Cleaning the structure
			Check and replacement of power system pieces
			Fuel, oil
		Electric	Cleaning the structure
			Checking and replacement of power system pieces
		Solar	Cleaning the structure
			Checking and replacement of power system pieces
		Wind	Cleaning the structure
			Checking and replacement of power system pieces

(*Continued*)

OPERATION AND MAINTENANCE COSTS			
ACTIVITY	**ITEM**	**SUB-ITEM**	**INPUT**
	Manual pumps	Suction plunger hand pump	Repair or replacement of pieces
		Direct action hand pump	Repair or replacement of pieces
		Deep-well piston pump	Repair or replacement of pieces
		Deep-well diaphragm pump	Repair or replacement of pieces
	Transport	Pipes	Repair or replacement of pieces
Water storage	Storage tank	Concrete-lined earthen reservoir	Cleaning, checking the valves, screens and lining
			Draining and disinfecting the reservoir
		Reinforced concrete reservoir	Cleaning, checking the valves and screens
			Draining and disinfecting the reservoir
		Elevated reservoir (tower)	Cleaning, checking the valves and screens
			Draining and disinfecting the reservoir
			Painting the tank and tower
		Ferro-cement tank	Cleaning, checking the valves and screens
			Draining and disinfecting the reservoir
Water distribution	Public standpipe	Connection	Replacement of pipe
		Supporting structure	Cleaning the structure, checking and repair of the platform, tap and valve
		Collecting spillage	Checking and repair of the slab, drain, soak-away
Water treatment	Household treatment	Slow sand filter	Cleaning or replacement
		Chlorination	Cleaning
	Central treatment	Storage and sedimentation	Regulation of inlet
			Flushing of deposited silt
			Repair or replacement of valve
		Up-flow roughing filters	Regulation of flow
			Cleaning of site
			Greasing of valve
			Repair or replacement of valve

OPERATION AND MAINTENANCE COSTS			
ACTIVITY	ITEM	SUB-ITEM	INPUT
		Slow sand filtration	Checking flow
			Cleaning site
			Scraping off sand, washing, drying and storage
			Re-sanding filter
			Repair or replacement of valve and metal tube
			Disinfecting filter outlet
		Chlorination in piped systems	Refilling chlorine tank
			Cleaning and replacement of chlorinator
			Checking and adjustment of doses
Equipment			
Water conveyance	Power pump	Axial flow pump	Repair or replacement
		Centrifugal pump	Repair or replacement
		Submersible pump	Repair or replacement
		Hydraulic ram pump	Repair or replacement
	Manual pump	Suction plunger hand pump	Repair or replacement
		Direct action hand pump	Repair or replacement
		Deep-well piston pump	Repair or replacement
		Deep-well diaphragm pump	Repair or replacement
Water distribution	Public standpipe	Connection	Replacement of the water meter
Water treatment	Household treatment	Slow sand filter	Replacement of pieces
	Central treatment	Storage and sedimentation	Repair or replacement of pieces
		Up-flow roughing filters	Repair or replacement of pieces
		Slow sand filtration	Repair or replacement of pieces
		Chlorine in piped systems	Repair or replacement of pieces
			Refilling chlorine tank

(*Continued*)

OPERATION AND MAINTENANCE COSTS			
ACTIVITY	**ITEM**	**SUB-ITEM**	**INPUT**
Labour			
Water collection	Well disinfection	Dug well	Skilled worker
	Well lining	Dug well	Repairs (semi-skilled worker)
			Extension (skilled worker)
			Cleaning (unskilled worker)
		Drilled well lining	Repairs (semi-skilled worker)
			Rehabilitation (skilled worker)
			Cleaning (unskilled worker)
Water conveyance	Power system	Diesel	Skilled worker
		Electric	Skilled worker
		Solar	Skilled worker
		Wind	Skilled worker
	Power pump	Axial flow pump	Plumber
		Centrifugal pump	Plumber
		Submersible pump	Plumber
		Diesel engine	Plumber
		Hydraulic ram pump	Plumber
	Transport	Pipes	Skilled worker
Water storage	Storage tank	Concrete-lined earthen reservoir	Semi-skilled worker (caretaker-inspection) and plumber
		Reinforced concrete reservoir	Semi-skilled worker (caretaker-inspection) and plumber
		Elevated reservoir (tower)	Semi-skilled worker (caretaker-inspection) and plumber
		Ferro-cement tank	Semi-skilled worker (caretaker-inspection) and plumber
Water distribution	Public standpipe	Connection	Plumber
		Supporting structure	Skilled worker
		Collecting spillage	Skilled worker
Water treatment	Household treatment	Heating	Family member
		Slow sand filter	Family member
		Chlorination	Family member

OPERATION AND MAINTENANCE COSTS			
ACTIVITY	**ITEM**	**SUB-ITEM**	**INPUT**
	Central treatment	Storage and sedimentation	Semi-skilled worker (caretaker-inspection) and plumber
		Up-flow roughing filters	Semi-skilled worker (caretaker-inspection) and plumber
		Slow sand filtration	Semi-skilled worker (caretaker-inspection) and plumber
		Chlorination in piped systems	Semi-skilled worker (caretaker-inspection) and plumber

III.3 TUBE WELL OR BOREHOLE

INVESTMENT COSTS			
ACTIVITY	**ITEM**	**SUB-ITEM**	**INPUT**
Materials			
Water collection	Well excavation and drilling	Hand-dug well	Various tools (shovel, pick, spud bar and others)
		Driven well	Sledgehammer, drive cap, metal well point
		Jetted well	Tripod and pulley, hollow drill (rod in section), pump, plastic pipe
		Bored well	Tripod, handle, coupling, auger iron, plastic or asbestos pipe
	Well lining	Dug well	Stone, brick or concrete (headwall), concrete rings, masonry with bricks or concrete blocks, clay, backfill (lining), gravel or prefabricated caisson intake
		Drilled well lining	Concrete, PVC pipe, gravel
		Fence	Concrete, wood
Water conveyance	Pumps	Manual pump	Bucket and rope system
		Diesel pump	Cables, superstructure (wood, slate, etc.)
		Electric pump	Cables, superstructure (wood, slate, etc.)
		Solar pump	Cables, superstructure (wood, slate, etc.)
		Wind pump	Cables, superstructure (wood, slate, etc.)
	Transport	Pipes	Concrete, PVC or others
Water storage	Storage tank	Concrete-lined earthen reservoir	Concrete, inlet and outlets, bituminous filling (or other waterproof materials) or Ferro-cement

(*Continued*)

INVESTMENT COSTS			
ACTIVITY	**ITEM**	**SUB-ITEM**	**INPUT**
		Reinforced concrete reservoir	Drain, concrete reinforced with steel bars or steel mesh, a roof of different materials (e.g. concrete reinforced), aeration pipe with a screen, a manhole, an inlet and outlets
		Elevated reservoir (tower)	For family use, the tank can be made of an oil-drum and the tower of bamboo
			For communal use, the tank can be made with galvanized steel and the tower of steel, wood or reinforced concrete. Several pipes and inspection entry are needed
		Ferro-cement tank	Aeration pipe, manhole, lid, wire screen mesh wrapped around the steel structure with a layer of cement, drainage and fence
Water distribution	Tap connection	Connection	Tap located in the lower part of the receiver tank
Water treatment	Household treatment	Slow sand filter	Sand and gravel
		Chlorination	Chlorine solutions
	Central treatment	Storage and sedimentation	Unlined pool or dirt with a lining of plastic (loam, clay, concrete, brick), chemicals
		Up-flow roughing filters	Sand, gravel, coconut husk fibre, bricks, concrete or Ferro-cement, PVC pipe, valve, plastic wire mesh
		Slow sand filtration	Sand, gravel, chemicals, valve, concrete, bricks, Ferro-cement, metal tube
		Chlorination in piped supply systems	Chlorine solution, valve, PVC tube
Equipment			
Water collection	Well excavation	Well drilling	Hydraulic rotary rigs or percussion rigs
Water conveyance	Manual pumps	Suction plunger hand-pump	Pump
		Direct action hand-pump	Pump
		Deep well piston-pump	Pump
		Deep well diaphragm-pump	Pump
	Power system	Diesel	Engine
		Electric	Engine, control panel
		Solar	Photovoltaic or solar cells

INVESTMENT COSTS			
ACTIVITY	**ITEM**	**SUB-ITEM**	**INPUT**
		Wind	Rotor fixed to a horizontal axis and mounted on a tower of steel. A gearbox is needed
	Power pump	Axial flow pump	Often a boat propeller is used to improvise an axial flow pump. It turns in cylindrical casing where fixed guide blades may be installed
		Centrifugal pump	Fast rotating impeller and casing
		Submersible pump	Pump and motor
		Hydraulic ram pump	Drive pipe, delivery valve, air chamber
Water treatment	Household treatment	Slow sand filter	Raw-water tank, filter tank and clean-water tank
	Central treatment	Chlorination in piped supply systems	Reservoir tank
Labour			
Water collection	Well excavation and drilling	Hand-dug well	Unskilled worker and skilled supervisor
		Driven well	Unskilled worker and skilled supervisor
		Jetted well	Unskilled worker and skilled supervisor
		Bored well	Unskilled worker and skilled supervisor
		Drilled well	Skilled worker
	Well lining	Dug well	Skilled worker or semi-skilled
		Drilled well lining	Skilled worker or semi-skilled
		Fence	Semi-skilled worker
Water conveyance	Manual pump	Rope and bucket	Unskilled worker and skilled supervisor
		Bucket pump	Semi-skilled worker and skilled supervisor
		Rope pump	Semi-skilled worker and skilled supervisor
		Suction plunger hand-pump	Plumber
		Direct action hand-pump	Plumber
		Deep well piston-pump	Plumber
		Deep well diaphragm-pump	Plumber
	Power system	Diesel	Skilled worker
		Electric	Skilled worker

(*Continued*)

INVESTMENT COSTS			
ACTIVITY	**ITEM**	**SUB-ITEM**	**INPUT**
		Solar	Skilled worker
		Wind	Skilled worker
	Power pump	Axial flow pump	Plumber
		Centrifugal pump	Plumber
		Submersible pump	Plumber
		Hydraulic ram pump	Plumber
	Transport	Pipes	Skilled worker
Water storage	Storage tank	Concrete-lined earthen reservoir	Skilled worker
		Reinforced concrete reservoir	Skilled worker
		Elevated reservoir (tower)	Skilled worker
		Ferro-cement tank	Skilled worker
Water distribution	Tap connection	Connection	Plumber
Water treatment	Household treatment	Slow sand filter	Member of the family
		Chlorination	Member of the family
	Central treatment	Storage and sedimentation	Plumber
		Up-flow roughing filters	Plumber
		Slow sand filtration	Plumber
		Chlorination in piped supply systems	Plumber

OPERATION AND MAINTENANCE COSTS			
ACTIVITY	**ITEM**	**SUB-ITEM**	**INPUT**
Materials			
Water collection	Well disinfection	Dug well	Chlorine (or other chemical product)
	Well lining	Dug well	Repairing the apron, headwall, drain

OPERATION AND MAINTENANCE COSTS			
ACTIVITY	**ITEM**	**SUB-ITEM**	**INPUT**
			Cleaning the apron, headwall, drain
			Extension of the well
		Drilled well lining	Repairing the apron, headwall, drain
			Cleaning the apron, headwall, drain
			Rehabilitation of the well (various technologies such as air-forced and water pumping, brushing and treatment with chemicals)
		Fence	Painting and replacement
Water conveyance	Manual pumps	Rope and bucket	Cleaning, greasing and replacement of pieces
		Bucket pump	Cleaning, greasing and replacement of pieces
		Rope pump	Cleaning, greasing and replacement of pieces and structure
		Suction plunger hand-pump	Cleaning and replacement of pieces
		Direct action hand-pump	Cleaning and replacement of pieces
		Deep well piston-pump	Cleaning and replacement of pieces
		Deep well diaphragm-pump	Cleaning and replacement of pieces
	Power system	Diesel	Cleaning the structure
			Checking and replacement of power system pieces
			Fuel, oil
		Electric	Cleaning the structure
			Checking and replacement of power system pieces
			Electricity
		Solar	Cleaning the structure
			Checking and replacement of power system pieces
		Wind	Cleaning the structure
			Checking and replacement of power system pieces
	Transport	Pipes	Replacement
Water storage	Storage tank	Concrete-lined earthen reservoir	Cleaning, checking the valves, screens and lining
			Draining and disinfecting the reservoir
		Reinforced concrete reservoir	Cleaning, checking the valves and screens
			Draining and disinfecting the reservoir

(*Continued*)

OPERATION AND MAINTENANCE COSTS			
ACTIVITY	**ITEM**	**SUB-ITEM**	**INPUT**
		Elevated reservoir (tower)	Cleaning, checking the valves and screens
			Draining and disinfecting the reservoir
			Painting the tank and tower
		Ferro-cement tank	Cleaning, checking the valves and screens
			Draining and disinfecting the reservoir
Water distribution	Tap connection	Connection	Replacement of taps
Water treatment	Household treatment	Slow sand filter	Cleaning or replacement
Equipment			
Water conveyance	Manual pump	Suction plunger hand-pump	Repair or replacement
		Direct action hand-pump	Repair or replacement
		Deep well piston-pump	Repair or replacement
		Deep well diaphragm-pump	Repair or replacement
	Power pump	Axial flow pump	Repair or replacement
		Centrifugal pump	Repair or replacement
		Submersible pump	Repair or replacement
		Hydraulic ram pump	Repair or replacement
Water treatment	Household treatment	Slow sand filter	Tank replacement
Labour			
Water collection	Well disinfection	Dug well	Skilled worker
	Well lining	Dug well	Repairs (semi-skilled worker)
			Extension (skilled worker)
			Cleaning (unskilled worker)
		Drilled well lining	Repairs (semi-skilled worker)
			Rehabilitation (skilled worker)
			Cleaning (unskilled worker)
		Fence	Semi-skilled worker

OPERATION AND MAINTENANCE COSTS			
ACTIVITY	**ITEM**	**SUB-ITEM**	**INPUT**
Water conveyance	Manual pump	Robe and bucket	Unskilled worker
		Bucket pump	Semi-skilled worker
		Suction plunger hand-pump	Plumber
		Direct action hand-pump	Plumber
		Deep well piston-pump	Plumber
		Deep well diaphragm-pump	Plumber
	Power system	Diesel	Skilled worker
		Electric	Skilled worker
		Solar	Skilled worker
		Wind	Skilled worker
	Power pump	Axial flow pump	Plumber
		Centrifugal pump	Plumber
		Submersible pump	Plumber
		Diesel engine	Plumber
		Hydraulic ram pump	Plumber
	Transport	Pipes	Semi-skilled worker
Water storage	Storage tank	Concrete-lined earthen reservoir	Semi-skilled worker (caretaker-inspection) and plumber
		Reinforced concrete reservoir	Semi-skilled worker (caretaker-inspection) and plumber
		Elevated steel reservoir (tower)	Semi-skilled worker (caretaker-inspection) and plumber
		Ferro-cement tank	Semi-skilled worker (caretaker-inspection) and plumber
Water distribution	Tap connection	Connection	Plumber
Water treatment	Household treatment	Heating	Household member
		Slow sand filter	Household member
		Chlorination	Household member

III.4 PROTECTED DUG WELL

INVESTMENT COSTS			
ACTIVITY	ITEM	SUB-ITEM	INPUT
Materials			
Water collection	Well excavation and drilling	Hand-dug well	Various tools (shovel, pick, spud bar and others)
		Driven well	Sledgehammer, drive cap, metal well point
		Jetted well	Tripod and pulley, hollow drill (rod in section), pump, plastic pipe
		Bored well	Tripod, handle, coupling, auger iron, plastic or asbestos pipe
	Well lining	Dug well	Stone, brick or concrete (headwall), concrete rings, masonry with bricks or concrete blocks, clay, backfill (lining), gravel or prefabricated caisson intake
		Drilled well lining	Concrete, PVC pipe, gravel
		Fence	Concrete, wood
Water conveyance	Pumps	Manual pump	Bucket and rope systems
		Diesel pump	Cables, superstructure (wood, slate, etc.)
		Electric pump	Cables, superstructure (wood, slate, etc.)
		Solar pump	Cables, superstructure (wood, slate, etc.)
		Wind pump	Cables, superstructure (wood, slate, etc.)
	Transport	Pipes	Concrete, PVC or other
Water storage	Storing tank	Concrete-lined earthen reservoir	Concrete, inlet and outlets, bituminous filling (or other waterproof materials) or Ferro-cement
		Reinforced concrete reservoir	Drain, concrete reinforced with steel bars or steel mesh, a roof of different materials (e.g. concrete reinforced), aeration pipe with a screen, a manhole, an inlet and outlets
		Elevated reservoir (tower)	For family use, the tank can be made of an oil-drum and the tower of bamboo For communal use, the tank can be made with galvanized steel and the tower of steel, wood or reinforced concrete. Several pipes and inspection entry are needed
		Ferro-cement tank	Aeration pipe, manhole, lid, wire screen mesh wrapped around the steel structure with a layer of cement, drainage and fence
Water distribution	Tap connection	Connection	Tap located in the lower part of the receiver tank

INVESTMENT COSTS			
ACTIVITY	**ITEM**	**SUB-ITEM**	**INPUT**
Water treatment	Household treatment	Slow sand filter	Sand and gravel
		Chlorination	Chlorine solutions
Equipment			
Water collection	Well excavation	Well drilling	Hydraulic rotary rigs or percussion rigs
Water conveyance	Manual pumps	Suction plunger hand-pump	Pump
		Direct action hand-pump	Pump
		Deep well piston-pump	Pump
		Deep well diaphragm-pump	Pump
	Power system	Diesel	Engine
		Electric	Engine, control panel
		Solar	Photovoltaic or solar cells
		Wind	Rotor fixed to a horizontal axis and mounted on a tower of steel. A gearbox is needed
	Power pump	Axial flow pump	Often a boat propeller is used to improvise an axial flow pump. It turns in s cylindrical casing where fixed guide blades may be installed
		Centrifugal pump	Fast rotating impeller and casing
		Submersible pump	Pump and motor
		Hydraulic ram pump	Drive pipe, delivery valve, air chamber
Water treatment	Household treatment	Slow sand filter	Raw water tank, filter tank and clean water tank
Labour			
Water collection	Well excavation and drilling	Hand-dug well	Unskilled worker and skilled supervisor
		Driven well	Unskilled worker and skilled supervisor
		Jetted well	Unskilled worker and skilled supervisor
		Bored well	Unskilled worker and skilled supervisor
		Drilled well	Skilled worker
	Well lining	Dug well	Skilled worker or semi-skilled
		Drilled well lining	Skilled worker or semi-skilled

(*Continued*)

INVESTMENT COSTS			
ACTIVITY	**ITEM**	**SUB-ITEM**	**INPUT**
		Fence	Semi-skilled worker
Water conveyance	Manual pump	Rope and bucket	Unskilled worker and skilled supervisor
		Bucket pump	Semi-skilled worker and skilled supervisor
		Rope pump	Semi-skilled worker and skilled supervisor
		Suction plunger hand-pump	Plumber
		Direct action hand-pump	Plumber
		Deep well piston-pump	Plumber
		Deep well diaphragm-pump	Plumber
	Power system	Diesel	Skilled worker
		Electric	Skilled worker
		Solar	Skilled worker
		Wind	Skilled worker
	Power pump	Axial flow pump	Plumber
		Centrifugal pump	Plumber
		Submersible pump	Plumber
		Hydraulic ram pump	Plumber
	Transport	Pipes	Skilled worker
Water storage	Storage tank	Concrete-lined earthen reservoir	Skilled worker
		Reinforced concrete reservoir	Skilled worker
		Elevated reservoir (tower)	Skilled worker
		Ferro-cement tank	Skilled worker
Water distribution	Tap connection	Connection	Plumber
Water treatment	Household treatment	Slow sand filter	Member of the family
		Chlorination	Member of the family

OPERATION AND MAINTENANCE COSTS			
ACTIVITY	**ITEM**	**SUB-ITEM**	**INPUT**
Materials			
Water collection	Well disinfection	Dug well	Chlorine (or other chemical product)
	Well lining	Dug well	Repairing the apron, headwall, drain
			Cleaning the apron, headwall, drain
			Extension of the well
		Drilled well lining	Repairing the apron, headwall, drain
			Cleaning the apron, headwall, drain
			Rehabilitation of the well (various technologies such as air-forced and water pumping, brushing and treatment with chemicals)
		Fence	Painting and replacement
Water conveyance	Manual pumps	Rope and bucket	Cleaning, greasing and replacement of pieces
		Bucket pump	Cleaning, greasing and replacement of pieces
		Rope pump	Cleaning, greasing and replacement of pieces and structure
		Suction plunger hand-pump	Cleaning and replacement of pieces
		Direct action hand-pump	Cleaning and replacement of pieces
		Deep well piston-pump	Cleaning and replacement of pieces
		Deep well diaphragm-pump	Cleaning and replacement of pieces
	Power system	Diesel	Cleaning the structure
			Checking and replacement of power system pieces
			Fuel, oil
		Electric	Cleaning the structure
			Checking and replacement of power system pieces
			Electricity
		Solar	Cleaning the structure
			Checking and replacement of power system pieces
		Wind	Cleaning the structure
			Checking and replacement of power system pieces
	Transport	Pipes	Replacement

(*Continued*)

OPERATION AND MAINTENANCE COSTS			
ACTIVITY	**ITEM**	**SUB-ITEM**	**INPUT**
Water storage	Storage tank	Concrete-lined earthen reservoir	Cleaning, checking the valves, screens and lining
			Draining and disinfecting the reservoir
		Reinforced concrete reservoir	Cleaning, checking the valves and screens
			Draining and disinfecting the reservoir
		Elevated reservoir (tower)	Cleaning, checking the valves and screens
			Draining and disinfecting the reservoir
			Painting the tank and tower
		Ferro-cement tank	Cleaning, checking the valves and screens
			Draining and disinfecting the reservoir
Water distribution	Tap connection	Connection	Replacement of taps
Water treatment	Household treatment	Slow sand filter	Cleaning or replacement
Equipment			
Water conveyance	Manual pump	Suction plunger hand-pump	Repair or replacement
		Direct action hand-pump	Repair or replacement
		Deep well piston-pump	Repair or replacement
		Deep well diaphragm-pump	Repair or replacement
	Power pump	Axial flow pump	Replacement of the pump
		Centrifugal pump	Replacement of the pump
		Submersible pump	Replacement of the pump
		Hydraulic ram pump	Replacement of the pump
Water treatment	Household treatment	Slow sand filter	Tank replacement
Labour			
Water collection	Well disinfection	Dug well	Skilled worker

OPERATION AND MAINTENANCE COSTS			
ACTIVITY	**ITEM**	**SUB-ITEM**	**INPUT**
	Well lining	Dug well	Repairs (semi-skilled worker)
			Extension (skilled worker)
			Cleaning (unskilled worker)
		Drilled well lining	Repairs (semi-skilled worker)
			Rehabilitation (skilled worker)
			Cleaning (unskilled worker)
		Fence	Semi-skilled worker
Water conveyance	Manual pump	Robe and bucket	Unskilled worker
		Bucket pump	Semi-skilled worker
		Suction plunger hand-pump	Plumber
		Direct action hand-pump	Plumber
		Deep well piston-pump	Plumber
		Deep well diaphragm-pump	Plumber
	Power system	Diesel	Skilled worker
		Electric	Skilled worker
		Solar	Skilled worker
		Wind	Skilled worker
	Pumping	Axial flow pump	Plumber
		Centrifugal pump	Plumber
		Submersible pump	Plumber
		Diesel engine	Plumber
		Hydraulic ram pump	Plumber
	Transport	Pipes	Semi-skilled worker
Water storage	Storage tank	Concrete-lined earthen reservoir	Semi-skilled worker (caretaker-inspection) and plumber
		Reinforced concrete reservoir	Semi-skilled worker (caretaker-inspection) and plumber
		Elevated steel reservoir (tower)	Semi-skilled worker (caretaker-inspection) and plumber

(*Continued*)

OPERATION AND MAINTENANCE COSTS			
ACTIVITY	**ITEM**	**SUB-ITEM**	**INPUT**
		Ferro-cement tank	Semi-skilled worker (caretaker-inspection) and plumber
Water distribution	Tap connection	Connection	Plumber
Water treatment	Household treatment	Heating	Household member
		Slow sand filter	Household member
		Chlorination	Household member

III.5 PROTECTED SPRING

INVESTMENT COSTS			
ACTIVITY	**ITEM**	**SUB-ITEM**	**INPUT**
Materials			
Water collection	Well excavation and drilling	Hand-dug well	Various tools (shovel, pick, spud bar and others)
		Driven well	Sledgehammer, drive cap, metal well point
		Jetted well	Tripod and pulley, hollow drill (rod in section), pump, plastic pipe
		Bored well	Tripod, handle, coupling, auger iron, plastic or asbestos pipe
	Well lining	Dug well	Stone, brick or concrete (headwall), concrete rings, masonry with bricks or concrete blocks, clay, backfill (lining), gravel or prefabricated caisson intake
		Drilled well lining	Concrete, PVC pipe, gravel
		Fence	Concrete, wood
Water conveyance	Pumps	Diesel pump	Cables, superstructure (wood, slate, etc.)
		Electric pump	Cables, superstructure (wood, slate, etc.)
		Solar pump	Cables, superstructure (wood, slate, etc.)
		Wind pump	Cables, superstructure (wood, slate, etc.)
	Transport	Pipes	Concrete, PVC or other
Water storage	Storage tank	Concrete-lined earthen reservoir	Concrete, inlet and outlets, bituminous filling (or other waterproof materials) or Ferro-cement

INVESTMENT COSTS			
ACTIVITY	**ITEM**	**SUB-ITEM**	**INPUT**
		Reinforced concrete reservoir	Drain, concrete reinforced with steel bars or steel mesh, a roof of different materials (e.g. reinforced concrete), aeration pipe with a screen, a manhole, an inlet and outlets
		Elevated reservoir (tower)	For family use, the tank can be made of an oil-drum and the tower of bamboo
			For communal use, the tank can be made with galvanized steel and the tower of steel, wood or reinforced concrete. Several pipes and inspection entry are needed
		Ferro-cement tank	Aeration pipe, manhole, lid, wire screen mesh wrapped around the steel structure with a layer of cement, drainage and fence
Water distribution	Tap connection	Connection (gravity, etc.)	Tap located in the lower part of the receiver tank
Water treatment	Household treatment	Slow sand filter	Sand and gravel
		Chlorination	Chlorine solutions
Equipment			
Water collection	Well excavation	Well drilling	Hydraulic rotary rigs or percussion rigs
Water conveyance	Power system	Diesel	Engine
		Electric	Engine, control panel
		Solar	Photovoltaic or solar cells
		Wind	Rotor fixed to a horizontal axis and mounted on a tower of steel. A gearbox is needed
Water treatment	Household treatment	Slow sand filter	Raw water tank, filter tank and clean water tank
Labour			
Water collection	Well excavation and drilling	Hand-dug well	Unskilled worker and skilled supervisor
		Driven well	Unskilled worker and skilled supervisor
		Jetted well	Unskilled worker and skilled supervisor
		Bored well	Unskilled worker and skilled supervisor
		Drilled well	Skilled worker
	Well lining	Dug well	Skilled worker or semi-skilled
		Drilled well lining	Skilled worker or semi-skilled
		Fence	Semi-skilled worker

(Continued)

INVESTMENT COSTS			
ACTIVITY	**ITEM**	**SUB-ITEM**	**INPUT**
Water conveyance	Power system	Diesel	Skilled worker
		Electric	Skilled worker
		Solar pump	Skilled worker
		Wind pump	Skilled worker
	Transport	Pipes	Skilled worker
Water storage	Storage tank	Concrete-lined earthen reservoir	Skilled worker
		Reinforced concrete reservoir	Skilled worker
		Elevated reservoir (tower)	Skilled worker
		Ferro-cement tank	Skilled worker
Water distribution	Tap connection	Connection	Plumber
Water treatment	Household treatment	Slow sand filter	Member of the family
		Chlorination	Member of the family

OPERATION AND MAINTENANCE COSTS			
ACTIVITY	**ITEM**	**SUB-ITEM**	**INPUT**
Materials			
Water collection	Well disinfection	Dug well	Chlorine (or other chemical product)
	Well lining	Dug well	Repairing the apron, headwall, drain
			Cleaning the apron, headwall, drain
			Extension of the well
		Drilled well lining	Repairing the apron, headwall, drain
			Cleaning the apron, headwall, drain
			Rehabilitation of the well (various technologies such as air-forced and water pumping, brushing and treatment with chemicals)
		Fence	Painting and replacement

OPERATION AND MAINTENANCE COSTS			
ACTIVITY	**ITEM**	**SUB-ITEM**	**INPUT**
Water conveyance	Power system	Diesel	Cleaning the structure
			Checking and replacement of power system pieces
			Fuel, oil
		Electric	Cleaning the structure
			Checking and replacement of power system pieces
			Electricity
		Solar	Cleaning the structure
			Checking and replacement of power system pieces
		Wind	Cleaning the structure
			Checking and replacement of power system pieces
	Transport	Pipes	Replacement
Water storage	Storage tank	Concrete-lined earthen reservoir	Cleaning, checking the valves, screens and lining
			Draining and disinfecting the reservoir
		Reinforced concrete reservoir	Cleaning, checking the valves and screens
			Draining and disinfecting the reservoir
		Elevated reservoir (tower)	Cleaning, checking the valves and screens
			Draining and disinfecting the reservoir
			Painting the tank and tower
		Ferro-cement tank	Cleaning, checking the valves and screens
			Draining and disinfecting the reservoir
Water distribution	Tap connection	Connection	Replacement of taps
Water treatment	Household treatment	Slow sand and filter	Cleaning or replacement
Equipment			
Water conveyance	Power system	Diesel power	Repair or replacement
		Electric power	Repair or replacement
		Solar power	Repair or replacement

(*Continued*)

OPERATION AND MAINTENANCE COSTS			
ACTIVITY	**ITEM**	**SUB-ITEM**	**INPUT**
		Wind power	Repair or replacement
Water treatment	Household treatment	Slow sand filter	Tank replacement
Labour			
Water collection	Well disinfection	Dug well	Skilled worker
	Well lining	Dug well	Repairs (semi-skilled worker)
			Extension (skilled worker)
			Cleaning (unskilled worker)
		Drilled well lining	Repairs (semi-skilled worker)
			Rehabilitation (skilled worker)
			Cleaning (unskilled worker)
		Fence	Semi-skilled worker
Water conveyance	Power system	Diesel	Skilled worker
		Electric	Skilled worker
		Solar	Skilled worker
		Wind	Skilled worker
	Transport	Pipes	Skilled worker
Water storage	Storing tank	Concrete-lined earthen reservoir	Semi-skilled worker (caretaker-inspection) and plumber
		Reinforced concrete reservoir	Semi-skilled worker (caretaker-inspection) and plumber
		Elevated steel reservoir (tower)	Semi-skilled worker (caretaker-inspection) and plumber
		Ferro-cement tank	Semi-skilled worker (caretaker-inspection) and plumber
Water distribution	Tap connection	Connection	plumber
Water treatment	Household treatment	Heating	Household member
		Slow sand filter	Household member
		Chlorination	Household member

III.6 RAINWATER COLLECTION

INVESTMENT COSTS			
ACTIVITY	**ITEM**	**SUB-ITEM**	**INPUT**
Materials			
Water collection	Down-gutter pipes	Wood	Wood gutter, tools such as hammer, nails
		Galvanized iron	Galvanized iron gutter, welting etc.
		PVC	PVC gutter, PVC adhesive, tools such as hammer, screwdriver
Water conveyance	Transport	Pipes	PVC pipes
Water storage	Storage tank	Reinforced concrete reservoir	Drain, concrete reinforced with steel bars or steel mesh, a roof of different materials (e.g. reinforced concrete), aeration pipe with a screen, a manhole, an inlet and outlets
		Ferro-cement tank	Aeration pipe, manhole, lid, wire screen mesh wrapped around the steel structure with a layer of cement, drainage and fence
Water distribution	Tap connection	Connection	Tap located in the lower part of the receiver tank
Water treatment	Household treatment	Heating	Water boiler
		Slow sand filter	Sand and gravel
		Chlorination	Chlorine solutions
Equipment Water treatment	Household treatment	Slow sand filter	Raw water tank, filter tank and clean water tank
Labour Water collection	Down-gutter pipe	Wood	Unskilled worker and skilled supervisor
		Galvanized iron	Unskilled worker and skilled supervisor
		PVC	Unskilled worker and skilled supervisor
Water conveyance	Transport	Pipes	Skilled worker
Water storage	Storage tank	Reinforced concrete reservoir	Skilled worker
		Ferro-cement tank	Skilled worker
Water distribution	Tap connection	Connection	Plumber
Water treatment	Household treatment	Heating	Household member
		Slow sand filter	Household member
		Chlorination	Household member

(*Continued*)

OPERATION AND MAINTENANCE COSTS			
ACTIVITY	**ITEM**	**SUB-ITEM**	**INPUT**
Materials			
Water collection	Down-gutter pipes	Wood, galvanized iron or PVC	Repairing or replacing the wood gutter
			Cleaning the gutter and down-pipes
Water conveyance	Transport	Pipes	Replacement
Water storage	Storage tank	Reinforced concrete reservoir	Cleaning, checking the valves and screens
			Draining and disinfecting the reservoir
		Ferro-cement tank	Cleaning, checking the valves and screens
			Draining and disinfecting the reservoir
Water distribution	Tap connection	Connection	Replacement of taps
Water treatment	Household treatment	Slow sand filter	Cleaning or replacement
Equipment Water treatment	Household treatment	Slow sand filter	Tank replacement
Labour Water collection	Down gutter pipes	Wood, galvanized iron or PVC	Repair or replacement (unskilled worker)
			Cleaning (unskilled worker)
Water conveyance	Transport	Pipes	Skilled worker
Water storage	Storage tank	Reinforced concrete reservoir	Semi-skilled worker (caretaker-inspection) and plumber
		Ferro-cement tank	Semi-skilled worker (caretaker-inspection) and plumber
Water treatment	Household treatment	Heating	Household member
		Slow sand filter	Household member
		Chlorination	Household member

Annex IV

Costing questionnaires

IV.1 INVESTMENT COSTS

IMPORTANT: Before filling in this table, you should read Annexes I and III. Annex I describes water supply technologies and Annex III lists the main components of the improved water supply technologies.

Fill in the following table. Make sure that you include **all local materials used**. If **imported** materials are used, please fill in a separate sheet. If appropriate, you may change or add items, sub-items or inputs, but please retain the same format.

INVESTMENT COSTS						
Item	**Sub-item**	**Input**	**Acquisition cost in local currency**	**Input**		**Date of acquisition (mm.yyyy)**
				Quantity	**Unit of measurement**	
Materials						
Water collection						
Down-gutter pipes (only for rooftop water harvesting)						
	Wood gutter	Wood gutter structure			m	
		Nails			kg	
		Other _______				
	Galvanized iron gutter	Galvanized iron gutter			m	
		Welding			pieces	
		Other _______				
	PVC gutter	PVC Gutter			m	

(*Continued*)

INVESTMENT COSTS						
Item	Sub-item	Input	Acquisition cost in local currency	Input		Date of acquisition (mm.yyyy)
				Quantity	Unit of measurement	
		PVC adhesive			litre	
		Other ______				
	Other ______	Other ______				
Catchments						
	Storage dams	Reinforced concrete			m^3	
		Non-reinforced concrete			m^3	
		Masonry			m^3	
		PVC			m	
		Other ______				
	Protected spring	Protective structure			m^3	
		Drain pipe			m	
		Seal (puddle clay)			m^3	
		PVC			m	
		Valve			pieces	
		Other ______				
	Other ______	Other ______				
Intakes						
	Protected side intake	Suction pipe			m	
		Reinforced concrete			m^3	
		Strainer filter (screen)			pieces	
		Other ______				
	Bottom river intake	Suction pipe			m	
		Reinforced concrete			m^3	
		Strainer filter (screen)			pieces	
		Stone pitching			pieces	
		Gabions			m^3	
		Other ______				
	Floating intake	Suction pipe			m	

INVESTMENT COSTS						
Item	Sub-item	Input	Acquisition cost in local currency	Input		Date of acquisition (mm.yyyy)
				Quantity	Unit of measurement	
		Floating pontoon			pieces	
		Strainer filter (screen)			pieces	
		Other ______				
	Sump intake	Suction pipe			m	
		Stone in gabions or timber crib			m^3	
		Sump as silt trap			pieces	
		Strainer filter (screen)			pieces	
		Other ______				
	Other ______	Other ______				
Well excavation and drilling						
	Hand-dug well	Shovel			pieces	
		Spud bar			pieces	
		Other ______				
	Driven well	Sledgehammer			pieces	
		Drive cap			pieces	
		Other ______				
	Jetted well	Pump			pieces	
		Plastic pipe			m	
		Other ______				
	Bored well	Auger iron			pieces	
		Plastic pipe			m	
		Other ______				
	Other ______	Other ______				
Well lining						
	Dug well	Cement			kg	
		Sand			kg	
		Gravel			m^3	
		Other ______				
	Drilled well lining	Cement			kg	
		Tube or borehole			m	
		Gravel			m^3	
		Sand			m^3	
		Other ______				

(*Continued*)

INVESTMENT COSTS						
Item	Sub-item	Input	Acquisition cost in local currency	Input		Date of acquisition (mm.yyyy)
				Quantity	Unit of measurement	
	Fence	Cement			kg	
		Wood			m^2	
		Other ______				
	Other ______	Other ______				
Water conveyance						
Pumps						
	Manual pump	Bucket			pieces	
		Rope			pieces	
		Bucket pump			pieces	
		Rope pump			pieces	
		Other ______			pieces	
	Diesel pump	Wood for superstructure			m^2	
		Slab for superstructure			m^2	
		Cables			m	
		Pipes			m	
		Other ______				
	Electric pump	Wood for superstructure			m^2	
		Slab for superstructure			m^2	
		Cables			m	
		Pipes			m	
		Other ______				
	Solar pump	Superstructure			pieces	
		Slab for superstructure			m^2	
		Cables			m	
		Pipes			m	
		Other ______				
	Wind pump	Superstructure			pieces	
		Slab for superstructure			m^2	
		Cables			m	

INVESTMENT COSTS						
Item	**Sub-item**	**Input**	**Acquisition cost in local currency**	**Input**		**Date of acquisition (mm.yyyy)**
				Quantity	**Unit of measurement**	
		Pipes			m	
		Other ______				
	Other ______	Other ______				
Transport						
	Manual transport	Bucket			pieces	
		Rope			m	
		Pulley			pieces	
		Windlass with hose			pieces	
		Other ______				
	Pipes	Cement			kg	
		Sand			m^3	
		PVC pipes			m	
		Other ______			m	
	Other ______	Other ______				
Water storage						
Storage tank						
	Concrete-lined earthen reservoir	Cement			kg	
		Sand			m^3	
		Gravel			m^3	
		Bitumen for lining			m^3	
		Pipes (inspection)			m	
		Other ______				
	Reinforced concrete reservoir	PVC tank or other			pieces	
		Steel mesh			kg	
		Concrete roof			m^2	
		PVC roof			m^2	
		Wood roof			m^2	
		Steel bars			kg	
		PVC aeration			m	
		Cement (drain)			kg	
		Reinforced concrete			m^3	

(*Continued*)

INVESTMENT COSTS						
Item	Sub-item	Input	Acquisition cost in local currency	Input		Date of acquisition (mm.yyyy)
				Quantity	Unit of measurement	
		Sand			m^3	
		Gravel			m^3	
		Pipes (inspection)			m	
		Screen			pieces	
		Manhole			pieces	
		Other ______				
	Elevated steel reservoir (tower)	Galvanized steel			kg	
		Steel			kg	
		Wood			m	
		Cement			kg	
		Reinforced concrete			m^3	
		Sand			m^3	
		Gravel			m^3	
		Pipes (inspection)			m	
		Other ______				
	Ferro-cement tank	PVC tank or other			pieces	
		PVC aeration			m	
		Wire screen			m	
		Wood for fence			m	
		Cement (drain)			kg	
		Sand			m^3	
		Gravel			m^3	
		Pipes (inspection)			m	
		Screen			pieces	
		Manhole			pieces	
		Other ______				
	Cement fence (Ferro-cement tank)	Cement			kg	
		Sand			m^3	
		Gravel			m^3	
		Other ______				
	Other ______	Other ______				

INVESTMENT COSTS						
Item	Sub-item	Input	Acquisition cost in local currency	Input		Date of acquisition (mm.yyyy)
				Quantity	Unit of measurement	
Water treatment						
Household treatment						
	Heating	Water boiler			pieces	
		Other ______				
	Chlorination	Chlorine solution			litre	
		Other ______				
	Iron filter	Ferro-cement iron filters			pieces	
		Sand			m^3	
		Gravel			m^3	
		Other ______				
	Slow sand filter	Sand			m^3	
		Gravel			m^3	
		Other ______				
	Other ______	Other ______				
Central treatment						
	Storage and sedimentation	Unlined pool			pieces	
		Reinforced concrete			m^3	
		Non-reinforced concrete			m^3	
		Other ______				
	Up-flow roughing filters	Sand			m^3	
		Gravel			m^3	
		Bricks			pieces	
		Concrete			m^3	
		PVC pipes			m	
		Valves			pieces	
		Other ______				
	Small filter	Charcoal			kg	
		Sand			m^3	
		Gravel			m^3	
		Other ______				

(Continued)

INVESTMENT COSTS						
Item	Sub-item	Input	Acquisition cost in local currency	Input		Date of acquisition (mm.yyyy)
				Quantity	Unit of measurement	
	Slow sand filtration	Sand			m^3	
		Gravel			m^3	
		Bricks			kg	
		Reinforced concrete			m^3	
		Non-reinforced concrete			m^3	
		Metal tube			m	
		Valves			m	
		Other ______				
	Chlorination in piped systems	Chlorine solution			litre	
		Valve			pieces	
		PVC pipe			m	
		Other ______				
	Other ______	Other ______				
Water distribution						
Tap connection						
	Connection	Tap			pieces	
		PVC pipe			m	
		Other ______				
	Other ______	Other ______				
Public standpipe						
	Connection	Tap			pieces	
		Suction hand pump			pieces	
		PVC pipe			m	
		Cement			kg	
		Sand			m^3	
		Other ______				
	Supporting structure	Wood			m	
		Brickwork			m^2	
		Dry stone masonry			m^2	
		Concrete			m^2	

INVESTMENT COSTS						
Item	**Sub-item**	**Input**	**Acquisition cost in local currency**	**Input**		**Date of acquisition (mm.yyyy)**
				Quantity	**Unit of measurement**	
		Globe or self-closing tap			pieces	
		Concrete slab or apron			kg	
		Regulating valve			pieces	
		Other ______				
	Cement fence (supporting)	Cement			kg	
		Sand			m^3	
		Gravel			m^3	
		Other ______				
	Wood fence (supporting)	Wood type (indicate: ______)			pieces	
		Paint			litre	
		Other ______				
	Spillage collector	Bricks			m^2	
		Strainer			pieces	
		Cement			kg	
		Sand			m^3	
		Gravel			m^3	
		Other ______				
	Other ______	Other ______				
Private connection						
	Connection	Tap			pieces	
		PVC pipe			m	
		Cement			kg	
		Sand			m^3	
		Valve			pieces	
		Other ______				
	Other ______	Other ______				

Fill in the following table. Make sure that you include **all local equipment used**. If **imported equipment** is used, please fill in a separate sheet. If appropriate, you may change or add items, sub-items or inputs, but please retain the same format.

INVESTMENT COSTS						
Item	**Sub-item**	**Input**	**Acquisition cost in local currency**	**Input**		**Date of acquisition (mm.yyyy)**
				Quantity	**Unit of measurement**	
Equipment						
Water collection						
Intakes						
	Protected side intake	Suction pump			pieces	
	Bottom river intake	Suction pump			pieces	
	Floating intake	Suction pump			pieces	
	Sump intake	Suction pump			pieces	
	Other ______	Other ______				
Well excavation						
	Well drilling	Hydraulics rotator rigs rented			hours	
		Percussion rigs rented			hours	
		Other ______				
	Other ______	Other ______				
Water conveyance						
Pumps						
	Manual pump	Suction plunger hand-pump			pieces	
		Direct action hand-pump			pieces	
		Deep well piston-pump			pieces	
		Deep well diaphragm-pump			pieces	
		Other ______				
	Diesel pump	Diesel engine			pieces	
		Other ______				
	Electric pump	Electric engine			pieces	
		Control panel			pieces	
		Other ______				
	Solar pump	Solar cells			pieces	
		Other ______				
	Wind pump	Wind tower			pieces	
		Other ______				

INVESTMENT COSTS						
Item	Sub-item	Input	Acquisition cost in local currency	Input		Date of acquisition (mm.yyyy)
				Quantity	Unit of measurement	
	Axial flow pump	Pump			pieces	
	Centrifugal pump	Pump			pieces	
	Submersible pump	Pump			pieces	
	Hydraulic ram pump	Pump			pieces	
	Other _______	Other _______				
Water treatment						
Household treatment						
	Slow and filter	Raw water tank			pieces	
		Filter tank			pieces	
		Clean water tank			pieces	
		Other _______				
	Other _______	Other _______				
Central treatment						
	Chlorination in piped systems	Reservoir tank			pieces	
		Other _______				
	Other _______	Other _______				
Water distribution						
Public standpipe						
	Trench	Bulldozer (rented, including operator)			hours	
		Compactor machine (rented, including operator)			hours	
		Other _______				
	Metering equipment	Water meter			pieces	
		Other _______				
	Other _______	Other _______				
Private connection						
	Trench	Bulldozer (rented, including operator)			hours	
		Compactor machine (rented, including operator)			hours	
		Other _______				
	Metering equipment	Water meter			pieces	
		Other _______				
	Other _______	Other _______				

Fill in the following table. Make sure that you include **all labour used**. If appropriate, you may change or add items, sub-items or inputs, but please retain the same format.

INVESTMENT COSTS						
Item	**Sub-item**	**Input**	**Input**		**Date of wage payment (mm.yyyy)**	**Total wage cost in local currency**
			Hourly wage in local currency	**Quantity in hours**		
Labour						
Water collection						
Down-gutter pipes (only for rooftop water harvesting)						
	Wood gutter	Unskilled worker				
		Skilled supervisor				
		Other ______				
	Galvanized iron gutter	Unskilled worker				
		Skilled supervisor				
		Other ______				
	PVC gutter	Unskilled worker				
		Skilled supervisor				
		Other ______				
	Other ______	Other ______				
Catchments						
	Storage dams	Skilled worker				
		Other ______				
	Protected spring	Skilled worker				
		Other ______				
	Other ______	Other ______				
Intakes						
	Protected side intake	Skilled worker				
		Other ______				
	Bottom river intake	Skilled worker				
		Other ______				
	Floating intake	Skilled worker				
		Other ______				
	Sump intake	Skilled worker				
		Other ______				
	Other ______	Other ______				

INVESTMENT COSTS						
Item	**Sub-item**	**Input**	**Input**		**Date of wage payment (mm.yyyy)**	**Total wage cost in local currency**
			Hourly wage in local currency	**Quantity in hours**		
Well excavation and drilling						
	Hand-dug well	Unskilled worker				
		Skilled supervisor				
	Driven well	Unskilled worker				
		Skilled supervisor				
	Jetted well	Unskilled worker				
		Skilled supervisor				
	Bored well	Unskilled worker				
		Skilled supervisor				
	Drilled well	Skilled worker				
		Skilled supervisor				
	Other _______	Other _______				
Well lining						
	Dug well	Skilled or semi-skilled worker				
		Skilled supervisor				
	Drilled well lining	Skilled or semi-skilled worker				
		Skilled supervisor				
	Fence	Semi-skilled worker				
		Other _______				
	Other _______	Other _______				
Water conveyance						
Manual pump						
	Rope and bucket	Unskilled worker				
		Skilled supervisor				
	Bucket pump	Semi-skilled worker				
		Skilled supervisor				
	Rope pump	Semi-skilled worker				
		Skilled supervisor				

(*Continued*)

INVESTMENT COSTS						
Item	Sub-item	Input	Input		Date of wage payment (mm.yyyy)	Total wage cost in local currency
			Hourly wage in local currency	Quantity in hours		
	Suction plunger hand-pump	Plumber				
	Direct action hand-pump	Plumber				
	Deep well piston-pump	Plumber				
	Deep well diaphragm-pump	Plumber				
	Other ______	Other ______				
Power pump						
	Axial flow pump	Plumber				
	Centrifugal pump	Plumber				
	Submersible pump	Plumber				
	Hydraulic ram pump	Plumber				
	Other ______	Other ______				
Power systems						
	Diesel, electric, solar and wind pump	Skilled worker				
	Other ______	Other ______				
Transport						
	Manual transport	Unskilled worker				
	Rope and bucket	Family member				
	Pipes	Plumber				
	Other ______	Other ______				
Water storage						
Storage tank						
	Concrete-lined earthen reservoir	Skilled worker				
	Reinforced concrete reservoir	Skilled worker				
	Elevated steel reservoir (tower)	Skilled worker				
	Ferro-cement tank	Skilled worker				
	Other ______	Other ______				

INVESTMENT COSTS						
Item	Sub-item	Input	Input		Date of wage payment (mm.yyyy)	Total wage cost in local currency
			Hourly wage in local currency	Quantity in hours		
Water treatment						
Household treatment						
	Heating	Family member				
	Slow sand filter	Family member				
	Chlorination	Family member				
	Other ______	Other ______				
Central treatment						
	Small filter	Semi-skilled worker				
	Storage and sedimentation	Plumber				
	Up-flow roughing filters	Plumber				
	Slow sand filtration	Plumber				
	Chlorination in piped systems	Plumber				
	Other ______	Other ______				
Water distribution						
Tap connection						
	Tap	Plumber				
	Other ______	Other ______				
Public standpipe						
	Trench	Unskilled worker				
	Connection	Skilled worker				
	Supporting	Unskilled worker				
	Cement fence	Unskilled worker				
	Wood fence	Unskilled worker				
	Collecting spillage	Skilled worker				
	Other ______	Other ______				
Private connection						
	Trench	Skilled worker				
	Connection	Plumber				
	Other ______	Other ______				

Fill in the following table. Make sure that you include **all other investment costs**. If appropriate, you may change or add items, sub-items or inputs, but please retain the same format.

OTHER INVESTMENT COSTS						
Item	**Sub-item**	**Input**	**Acquisition cost in local currency**	**Assessment**		**Date of acquisition (mm.yyyy)**
				Criterion	**Quantity**	
Other investment costs						
	Preliminary studies					
		Studies				
		Other ______				
	Administration costs	Management				
		Administrative services				
		Other ______				
	Promotion and training	In administration				
		In maintenance				
		In operation				
		Other ______				
	Education	Health education				
		Other ______				
	Other ______	Other ______				

Fill in the following table. Make sure that you include **all incidental investment costs**. If appropriate, you may change or add items, sub-items or inputs, but please retain the same format.

INCIDENTAL INVESTMENT COSTS						
Item	**Sub-item**	**Input**	**Acquisition cost in local currency**	**Assessment**		**Date of acquisition (mm.yyyy)**
				Criterion	**Quantity**	
Incidental investment costs						
	Work supervision					
	Engineering studies					
	Other contingencies					
	Other ______					

IV.2 MAINTENANCE COSTS

Fill in the following table. Make sure that you include **all local materials used in maintenance at design capacity**. If **imported materials** are used in maintenance, please fill in a separate sheet. If appropriate, you may change or add items, sub-items or inputs, but please retain the same format.

MAINTENANCE COSTS								
Item	Sub-item	Input	Periodicity in years round	Fix or variable cost	Cost at design capacity in local currency (mm.yyyy)	Input		Average annual growth rate of real unit cost (%)
						Quantity	Unit of measure-ment	
Materials								
Water collection								
Down-gutter pipes (only for rooftop water harvesting)								
	Wood gutter	Replacement of pieces					pieces	
		Cleaning					pieces	
		Other ______						
	Galvanized iron gutter	Replacement of pieces					pieces	
		Cleaning					pieces	
		Other ______						
	PVC gutter	Replacement of pieces					pieces	
		Cleaning					pieces	
		Other ______						
	Other ______	Other ______						
Catchments								
	Storage dams	Replacement of pieces					pieces	
		Structural repairs					m^2	
		Cleaning					pieces	
		Other ______						
	Protected spring	Replacement of pieces					pieces	
		Structural repairs					m^2	
		Cleaning					pieces	
		Other ______						
	Other ______	Other ______						

(Continued)

MAINTENANCE COSTS								
Item	**Sub-item**	**Input**	**Periodicity in years round**	**Fix or variable cost**	**Cost at design capacity in local currency (mm.yyyy)**	**Input**		**Average annual growth rate of real unit cost (%)**
						Quantity	**Unit of measure-ment**	
Intakes								
	Protected side intake	Replacement of pieces					pieces	
		Painting					litre	
		Structural repairs					m^2	
		Cleaning and desalting					pieces	
		Other ______						
	Bottom river intake	Replacement of pieces					pieces	
		Painting					litre	
		Structural repairs					m^2	
		Cleaning and desalting					pieces	
		Other ______						
	Floating intake	Replacement of pieces					pieces	
		Pontoon repairs					pieces	
		Painting					litre	
		Cleaning and desalting					pieces	
		Other ______						
	Sump intake	Replacement of pieces					pieces	
		Structural repairs					m^2	
		Cleaning and desalting					pieces	
		Other ______						
	Other ______	Other ______						
Well disinfection								
	Dug well	Chlorine					tablets	
		Sodium hypochlorite					litre	
		Other ______						
	Other ______	Other ______						

MAINTENANCE COSTS								
Item	Sub-item	Input	Periodicity in years round	Fix or variable cost	Cost at design capacity in local currency (mm.yyyy)	Input		Average annual growth rate of real unit cost (%)
						Quantity	Unit of measure-ment	
Well lining								
	Dug well	Replacement of pieces					pieces	
		Extension of well					m^3	
		Other ______						
	Drilled well lining	Replacement of pieces					pieces	
		Extension of well					m^3	
		Other ______						
	Fence	Replacement of pieces					pieces	
		Paint					litre	
	Other ______	Other ______						
Water conveyance								
Manual pumps								
	Rope and bucket	Replacement of pieces					pieces	
		Grease for bearing					pieces	
		Bucket					pieces	
		Rope					m	
		Other ______						
	Bucket pump	Replacement of pieces					pieces	
		Grease for bearing					pieces	
		Other ______						
	Rope pump	Replacement of pieces					pieces	
		Repairs of structure					m^2	
		Grease for bearing					pieces	
		Bucket					pieces	
		Rope					m	

(Continued)

MAINTENANCE COSTS								
Item	**Sub-item**	**Input**	**Periodicity in years round**	**Fix or variable cost**	**Cost at design capacity in local currency (mm.yyyy)**	**Input**		**Average annual growth rate of real unit cost (%)**
						Quantity	**Unit of measurement**	
		Paint					litre	
		Other ______						
	Suction plunger hand-pump	Replacement of pieces					pieces	
		Other ______						
	Direct action hand-pump	Replacement of pieces					pieces	
		Other ______						
	Deep well piston-pump	Replacement of pieces					pieces	
		Other ______						
	Deep well diaphragm-pump	Replacement of pieces					pieces	
		Other ______						
	Other ______	Other ______						
Power system								
	Diesel pump	Replacement of pieces					pieces	
		Other ______						
	Electric pump	Replacement of pieces					pieces	
		Other ______						
	Solar pump	Replacement of pieces					pieces	
		Other ______						
	Wind pump	Replacement of pieces					pieces	
		Other ______						
	Other ______	Other ______						
Transport								
	Manual transport	Replacement of bucket					pieces	
		Replacement of rope					m	
		Other ______						

MAINTENANCE COSTS								
Item	**Sub-item**	**Input**	**Periodicity in years round**	**Fix or variable cost**	**Cost at design capacity in local currency (mm.yyyy)**	**Input**		**Average annual growth rate of real unit cost (%)**
						Quantity	**Unit of measurement**	
	Pipes	Replacement of pipes					m	
		Cement					kg	
		Sand					m^3	
		Other _____						
	Other _____	Other _____						
Water storage								
Storage tank								
	Concrete-lined earthen reservoir	Replacement of pieces					pieces	
		Valves					pieces	
		Screens					pieces	
		Concrete					kg	
		Other _____						
	Reinforced concrete reservoir	Replacement of pieces					pieces	
		Valves					pieces	
		Brushes					pieces	
		Sponges					pieces	
		Screens					pieces	
		Other _____						
	Elevated steel reservoir (tower)	Replacement of pieces					pieces	
		Valves					pieces	
		Screens					pieces	
		Other _____						
	Ferro-cement tank	Replacement of pieces					pieces	
		Valves					pieces	
		Brushes					pieces	
		Sponges					pieces	
		Chemicals for disinfection					litre	
		Screens					pieces	

(*Continued*)

MAINTENANCE COSTS								
Item	**Sub-item**	**Input**	**Periodicity in years round**	**Fix or variable cost**	**Cost at design capacity in local currency (mm.yyyy)**	**Input**		**Average annual growth rate of real unit cost (%)**
						Quantity	**Unit of measurement**	
		Other ______						
	Other ______	Other ______						
Water treatment								
Household treatment								
	Heating	Replacement of water boiler					pieces	
		Other ______						
	Slow sand filter	Replacement of pieces					pieces	
		Other ______						
	Iron filter	Replacement of pieces					pieces	
		Other ______						
	Chlorination	Replacement of pieces					pieces	
		Replacement of tablets					tablets	
		Replacement of chlorine					litre	
		Other ______						
	Other ______	Other ______						
Central treatment								
	Storage and sedimentation	Replacement of pieces					pieces	
		Valves					pieces	
		Other ______						
	Up-flow roughing filters	Replacement of pieces					pieces	
		Grease for valve					pieces	
		Valves					pieces	
		Other ______						
	Slow sand filtration	Replacement of pieces					pieces	
		Grease for valve					pieces	

MAINTENANCE COSTS								
Item	Sub-item	Input	Periodicity in years round	Fix or variable cost	Cost at design capacity in local currency (mm.yyyy)	Input		Average annual growth rate of real unit cost (%)
						Quantity	Unit of measure-ment	
		Sand					m^3	
		Filter					pieces	
		Valves					pieces	
		Other ______						
	Small filter	Replacement of pieces					pieces	
		Other ______						
	Chlorine in piped systems	Refill chlorine tank					litre	
		Replacement of pieces					pieces	
		Other ______						
	Other ______	Other ______						
Water distribution								
Tap connection								
	Connection	Replacement of tap					pieces	
		Other ______						
	Other ______	Other ______						
Public standpipe								
	Standpipe connection	Replacement of tap					pieces	
		Replacements of pipes					m	
		Drain					m	
		Strainer					pieces	
		Cement					kg	
		Sand					m^3	
		Other ______						
	Fence	Cement					kg	
		Sand					m^3	
		Gravel					m^3	
		Wood					kg	
		Paint					litre	
		Other ______						
	Other ______	Other ______						

(*Continued*)

MAINTENANCE COSTS								
Item	**Sub-item**	**Input**	**Periodicity in years round**	**Fix or variable cost**	**Cost at design capacity in local currency (mm.yyyy)**	**Input**		**Average annual growth rate of real unit cost (%)**
						Quantity	**Unit of measure-ment**	
Private connection								
	Connection	Replacement of pieces					pieces	
		Replacement of tap					pieces	
		Replacement of pipes					m	
		Drain					m	
		Strainer					pieces	
		Cement					kg	
		Sand					m^3	
		Other ______						
	Other ______	Other ______						

Fill in the following table. Make sure that you include all the types of **local equipment used in maintenance at design capacity**. If **imported equipment** is used in maintenance, please fill in a separate sheet. If appropriate, you may change or add items, sub-items or inputs, but please retain the same format.

MAINTENANCE COSTS								
Item	**Sub-item**	**Input**	**Periodicity in years round**	**Fix or variable cost**	**Cost at design capacity in local currency (mm.yyyy)**	**Input**		**Average annual growth rate of real unit cost (%)**
						Quantity	**Unit of measure-ment**	
Equipment								
Water collection								
Intakes								
	Suction pump	Replacement or repairs					pieces	
		Other ______						
	Other ______	Other ______						

MAINTENANCE COSTS								
Item	Sub-item	Input	Periodicity in years round	Fix or variable cost	Cost at design capacity in local currency (mm.yyyy)	Input		Average annual growth rate of real unit cost (%)
						Quantity	Unit of measure-ment	
Water conveyance								
Manual pump								
		Suction plunger hand-pump	Replacement or repairs				pieces	
			Other ______					
		Direct action hand-pump	Replacement or repairs				pieces	
			Other ______					
		Deep well piston-pump	Replacement or repairs				pieces	
			Other ______					
		Deep well diaphragm-pump	Replacement or repairs				pieces	
			Other ______					
		Other ______	Other ______					
Power pump								
		Axial flow pump	Pump replacement or repairs				pieces	
			Other ______					
		Centrifugal pump	Pump replacement or repairs				pieces	
			Other ______					
		Submersible pump	Pump replacement or repairs				pieces	
			Other ______					
		Hydraulic ram pump	Pump replacement or repairs				pieces	
			Other ______					
		Other ______	Other ______					
Power system								
		Diesel power	Power replacement or repairs				pieces	
			Other ______					
		Electric power	Power replacement or repairs				pieces	
			Other ______					

(*Continued*)

MAINTENANCE COSTS								
Item	**Sub-item**	**Input**	**Periodicity in years round**	**Fix or variable cost**	**Cost at design capacity in local currency (mm.yyyy)**	**Input**		**Average annual growth rate of real unit cost (%)**
						Quantity	**Unit of measure-ment**	
		Solar power	Power replacement or repairs				pieces	
			Other ______					
		Wind power	Power replacement or repairs				pieces	
			Other ______					
		Other ______	Other ______					
Water treatment								
Household treatment								
		Slow sand filter	Raw water tank replacement				pieces	
			Filter tank replacement				pieces	
			Clean water tank replacement				pieces	
			Other ______					
		Other ______	Other ______					
Central treatment								
		Chlorination in piped systems	Reservoir tank replacement or repair				pieces	
			Other ______					
		Other ______	Other ______					
Water distribution								
Tap connection								
		Connection	Replacement of water meter				pieces	
			Other ______					
Private connection								
		Connection	Replacement of water meter				pieces	
			Other ______					
		Other ______	Other ______					

Fill in the following table. Make sure that you include **all labour used in maintenance at design capacity**. If appropriate, you may change or add items, sub-items or inputs, but please retain the same format.

MAINTENANCE COSTS								
Item	**Sub-item**	**Input**	**Periodicity in years round**	**Fix or variable cost**	**Cost at design capacity in local currency (mm.yyyy)**	**Input**		**Average annual growth rate of real unit cost (%)**
						Hourly wage in local currency	**Quantity in hours**	
Labour								
Water collection								
Down-gutter pipes (only for rooftop water harvesting)								
	Wood gutter	Repair or replacement – Unskilled worker						
		Cleaning – Unskilled worker						
		Other ______						
	Galvanized iron gutter	Repair or replacement - Semi-skilled worker						
		Cleaning – Unskilled worker						
		Other ______						
	PVC gutter	Repair or replacement - Semi-skilled worker						
		Cleaning – Unskilled worker						
		Other ______						
	Other ______	Other ______						
Catchments								
	Storage dams	Repair or replacement – Skilled worker						
		Cleaning – Unskilled worker						
		Other ______						
	Protected spring	Repair or replacement – Skilled worker						

(Continued)

MAINTENANCE COSTS								
Item	**Sub-item**	**Input**	**Periodicity in years round**	**Fix or variable cost**	**Cost at design capacity in local currency (mm.yyyy)**	**Input**		**Average annual growth rate of real unit cost (%)**
						Hourly wage in local currency	**Quantity in hours**	
		Cleaning – Unskilled worker						
		Other _______						
	Other _______	Other _______						
Intakes								
	Protected side intake	Repair or replacement – Skilled worker						
		Cleaning – Unskilled worker						
		Other _______						
	Bottom river intake	Repair or replacement – Skilled worker						
		Cleaning – Unskilled worker						
		Other _______						
	Floating intake	Repair or replacement – Skilled worker						
		Cleaning – Unskilled worker						
		Other _______						
	Sump intake	Repair or replacement – Skilled worker						
		Cleaning – Unskilled worker						
		Other _______						
	Other _______	Other _______						
Well disinfection								
	Dug well	Unskilled worker						
		Other _______						

MAINTENANCE COSTS								
Item	Sub-item	Input	Periodicity in years round	Fix or variable cost	Cost at design capacity in local currency (mm.yyyy)	Input		Average annual growth rate of real unit cost (%)
						Hourly wage in local currency	Quantity in hours	
Well lining								
	Dug well	Repairs - Semi-skilled worker						
		Extension – Skilled worker						
		Cleaning – Unskilled worker						
		Other _______						
	Drilled well lining	Repairs - Semi-skilled worker						
		Rehabilitation – Skilled worker						
		Cleaning – Unskilled worker						
		Other _______						
	Fence	Repairs - Semi-skilled worker						
		Other _______						
	Other _______	Other _______						
Water conveyance								
Manual pump								
	Rope and bucket	Unskilled worker						
		Other _______						
	Bucket pump	Semi-skilled worker						
		Other _______						
	Suction plunger hand-pump	Plumber						
		Other _______						
	Direct action hand-pump	Plumber						
		Other _______						

(Continued)

MAINTENANCE COSTS								
Item	**Sub-item**	**Input**	**Periodicity in years round**	**Fix or variable cost**	**Cost at design capacity in local currency (mm.yyyy)**	**Input**		**Average annual growth rate of real unit cost (%)**
						Hourly wage in local currency	**Quantity in hours**	
	Deep well piston-pump	Plumber						
		Other ______						
	Deep well diaphragm-pump	Plumber						
		Other ______						
	Other ______	Other ______						
Power pump								
	Axial flow pump	Plumber						
		Other ______						
	Centrifugal pump	Plumber						
		Other ______						
	Submersible pump	Plumber						
		Other ______						
	Hydraulic ram pump	Plumber						
		Other ______						
	Other ______	Other ______						
Power system								
	Diesel pump	Skilled worker						
		Other ______						
	Electric pump	Skilled worker						
		Other ______						
	Solar pump	Skilled worker						
		Other ______						
	Wind pump	Skilled worker						
		Other ______						
	Other ______	Other______						
Transport								
	Rope and bucket	Family member						
		Other ______						

MAINTENANCE COSTS								
Item	**Sub-item**	**Input**	**Periodicity in years round**	**Fix or variable cost**	**Cost at design capacity in local currency (mm.yyyy)**	**Input**		**Average annual growth rate of real unit cost (%)**
						Hourly wage in local currency	**Quantity in hours**	
	Pipes	Skilled worker						
		Other ______						
	Other ______	Other ______						
Water storage								
Storage tank								
	Concrete-lined earthen reservoir	Semi-skilled worker (caretaker-inspection)						
		Plumber						
		Other ______						
	Reinforced concrete reservoir	Semi-skilled worker (caretaker-inspection)						
		Plumber						
		Other ______						
	Elevated steel reservoir (tower)	Semi-skilled worker (caretaker-inspection)						
		Plumber						
		Other ______						
	Ferro-cement tank	Semi-skilled worker (caretaker-inspection)						
		Plumber						
		Other ______						
	Other ______	Other ______						
Water treatment								
Household treatment								
	Heating	Family member						
		Other ______						
	Slow sand filter	Family member						
		Other ______						

(Continued)

MAINTENANCE COSTS								
Item	**Sub-item**	**Input**	**Periodicity in years round**	**Fix or variable cost**	**Cost at design capacity in local currency (mm.yyyy)**	**Input**		**Average annual growth rate of real unit cost (%)**
						Hourly wage in local currency	**Quantity in hours**	
	Chlorination	Family member						
		Other ______						
	Iron filter	Family member						
		Other ______						
	Other ______	Other ______						
Central treatment								
	Storage and sedimentation	Skilled worker (caretaker-inspection)						
		Other ______						
	Up-flow roughing filters	Skilled worker (caretaker-inspection)						
		Other ______						
	Slow sand filtration	Skilled worker (caretaker-inspection)						
		Other ______						
	Small filter	Skilled worker (caretaker-inspection)						
		Other ______						
	Chlorination in piped systems	Skilled worker (caretaker-inspection)						
		Other ______						
	Other ______	Other ______						
Water distribution								
Tap connection								
	Connection	Plumber						
		Other ______						
	Other ______	Other ______						
Public standpipe								
	Connection	Plumber						
		Other ______						

MAINTENANCE COSTS								
Item	**Sub-item**	**Input**	**Periodicity in years round**	**Fix or variable cost**	**Cost at design capacity in local currency (mm.yyyy)**	**Input**		**Average annual growth rate of real unit cost (%)**
						Hourly wage in local currency	**Quantity in hours**	
	Fence	Unskilled worker						
		Other _______						
	Drain	Skilled worker						
		Other _______						
	Other_______	Other _______						
Private connection								
	Connection	Plumber						
		Other _______						
	Other _______	Other _______						

IV.3 OPERATION COSTS

Fill in the following table. Make sure that you include **all local materials used in operation at design capacity**. If **imported materials** are used in operations, please fill in a separate sheet. If appropriate, you may change or add items, sub-items or inputs, but please retain the same format.

OPERATION COSTS								
Item	**Sub-item**	**Input**	**Periodicity in years round**	**Fix or variable cost**	**Cost at design capacity in local currency (mm.yyyy)**	**Input**		**Average annual growth rate of real unit cost (%)**
						Quantity	**Unit of measure-ment**	
Materials								
Water collection								
Well disinfection								
	Dug well	Chlorine					tablets	
		Sodium hypochlorite					litre	
		Other _______						
	Other _______	Other _______						

(*Continued*)

OPERATION COSTS								
Item	**Sub-item**	**Input**	**Periodicity in years round**	**Fix or variable cost**	**Cost at design capacity in local currency (mm.yyyy)**	**Input**		**Average annual growth rate of real unit cost (%)**
						Quantity	**Unit of measure-ment**	
Water storage								
Storage tank								
	Tank	Chlorine					tablets	
		Sodium hypochlorite					litre	
		Other ______						
	Other ______	Other ______						
Water treatment								
Household treatment								
	Heating	Fuel					litre	
		Wood					kg	
		Other ______						
	Slow sand filter	Chlorine					tablets	
		Sodium hypochlorite					litre	
		Other ______						
	Chlorination	Chlorine					tablets	
		Sodium hypochlorite					litre	
		Other ______						
	Other ______	Other ______						
Central treatment								
	Storage and sedimentation	Chlorine					tablets	
		Sodium hypochlorite					litre	
		Other ______						
	Up-flow roughing filters	Filter					pieces	
	Slow sand filtration	Sand					m^3	
		Filter					pieces	
		Other ______						
	Small filter	Chlorine					tablets	
		Sodium hypochlorite					litre	
		Other ______						

OPERATION COSTS								
Item	**Sub-item**	**Input**	**Periodicity in years round**	**Fix or variable cost**	**Cost at design capacity in local currency (mm.yyyy)**	**Input**		**Average annual growth rate of real unit cost (%)**
						Quantity	**Unit of measure-ment**	
	Chlorine in piped systems	Chlorine					tablets	
		Sodium hypochlorite					litre	
		Other _______						
	Other _______	Other _______						

Fill in the following table. Make sure that you include **all types of local power services used in operation at design capacity**. If **imported power services** are used in operations, please fill in a separate sheet. If appropriate, you may change or add items, sub-items or inputs, but please retain the same format.

OPERATION COSTS								
Item	**Sub-item**	**Input**	**Periodicity in years round**	**Fix or variable cost**	**Cost at design capacity in local currency (mm.yyyy)**	**Input**		**Average annual growth rate of real unit cost (%)**
						Quantity	**Unit of measure-ment**	
Power services								
Water collection								
Intakes								
	Suction pump	Diesel					litre	
		Electricity					kWh	
		Other _______						
	Other _______	Other _______						
Other	Other _______	Other _______						
Water conveyance								
Pumps								
	Diesel pump	Oil					litre	
		Fuel					litre	
		Other _______						
	Electric pump	Electricity					kWh	
		Other _______						
	Other _______	Other _______						

(Continued)

OPERATION COSTS								
Item	**Sub-item**	**Input**	**Periodicity in years round**	**Fix or variable cost**	**Cost at design capacity in local currency (mm.yyyy)**	**Input**		**Average annual growth rate of real unit cost (%)**
						Quantity	**Unit of measure-ment**	
Transport								
	Diesel pump	Oil					litre	
		Fuel					litre	
		Other ______						
	Electric pump	Electricity					kWh	
		Other ______						
	Other ______	Other ______						
Other	Other ______	Other ______						
Water treatment								
Central treatment								
	Diesel pump	Oil					litre	
		Fuel					litre	
		Other ______						
	Electric pump	Electricity					kWh	
		Other ______						
	Other ______	Other ______						
Other	Other ______	Other ______						
Water distribution								
Private connection								
	Diesel pump	Oil					litre	
		Fuel					litre	
		Other ______						
	Electric pump	Electricity					kWh	
		Other ______						
	Other ______	Other ______						
Other	Other ______	Other ______						

Fill in the following table. Make sure that you include **all labour used in operation at design capacity**. If appropriate, you may change or add items, sub-items or inputs, but please retain the same format.

OPERATION COSTS								
Item	Sub-item	Input	Periodicity in years round	Fix or variable cost	Cost at design capacity in local currency (mm.yyyy)	Input		Average annual growth rate of real unit cost (%)
						Hourly wage in local currency	Quantity in hours	
Labour								
Water collection								
Catchments								
	Storage dams	Caretaker						
		Other _______						
	Protected spring	Caretaker						
		Other _______						
	Other _______	Other _______						
Intakes								
	Protected side intake	Caretaker						
		Other _______						
	Bottom river intake	Caretaker						
		Other _______						
	Floating intake	Caretaker						
		Other _______						
	Sump intake	Caretaker						
		Other _______						
	Other _______	Other _______						
Well disinfection								
	Dug well	Unskilled worker						
		Other _______						
	Other _______	Other _______						
Water conveyance								
Pump								
	Manual pump	Unskilled worker						
		Other _______						
	Axial flow pump	Semi-skilled worker						
		Other _______						

(*Continued*)

OPERATION COSTS								
Item	**Sub-item**	**Input**	**Periodicity in years round**	**Fix or variable cost**	**Cost at design capacity in local currency (mm.yyyy)**	**Input**		**Average annual growth rate of real unit cost (%)**
						Hourly wage in local currency	**Quantity in hours**	
	Centrifugal pump	Semi-skilled worker						
		Other ______						
	Submersible pump	Semi-skilled worker						
		Other ______						
	Hydraulic ram pump	Semi-skilled worker						
		Other ______						
	Other ______	Other ______						
Power system								
	Diesel pump	Semi-skilled worker						
		Other ______						
	Electric pump	Semi-skilled worker						
		Other ______						
	Solar pump	Semi-skilled worker						
		Other ______						
	Wind pump	Semi-skilled worker						
		Other ______						
	Other ______	Other ______						
Transport								
	Pipes	Semi-skilled worker						
		Other ______						
	Other ______	Other ______						
Water storage								
Storage tank								
	Tank	Semi-skilled worker						
		Other ______						
	Other ______	Other ______						

OPERATION COSTS								
Item	**Sub-item**	**Input**	**Periodicity in years round**	**Fix or variable cost**	**Cost at design capacity in local currency (mm.yyyy)**	**Input**		**Average annual growth rate of real unit cost (%)**
						Hourly wage in local currency	**Quantity in hours**	
Water treatment								
Household treatment								
	Heating	Family member						
		Other ______						
	Slow sand filter	Family member						
		Other ______						
	Iron filter	Family member						
		Other ______						
	Chlorination	Family member						
		Other ______						
	Other ______	Other ______						
Central treatment								
	Storage and sedimentation	Semi-skilled worker						
		Other ______						
	Up-flow roughing filters	Semi-skilled worker						
		Other ______						
	Slow sand filtration	Semi-skilled worker						
		Other ______						
	Small filter	Semi-skilled worker						
		Other ______						
	Chlorine in piped systems	Semi-skilled worker						
		Other ______						
	Other ______	Other ______						
Water distribution								
Tap connection								
	Connection	Skilled worker						
		Other ______						
	Other ______	Other ______						

(Continued)

OPERATION COSTS								
Item	Sub-item	Input	Periodicity in years round	Fix or variable cost	Cost at design capacity in local currency (mm.yyyy)	Input		Average annual growth rate of real unit cost (%)
						Hourly wage in local currency	Quantity in hours	
Public standpipe								
	Standpipe connection	Skilled worker						
		Other ______						
	Other ______	Other ______						
Private connection								
	Connection	Skilled worker						
		Other ______						
	Other ______	Other ______						

IV.4 OTHER RECURRENT COSTS

Fill in the following table. Make sure that you include **all other recurrent costs used at design capacity**. If appropriate, you may change or add items, sub-items or inputs, but please retain the same format.

OTHER RECURRENT COSTS								
Item	Sub-item	Input	Periodicity in years round	Fix or variable cost	Cost at design capacity in local currency (mm.yyyy)	Assessment		Average annual growth rate of real unit cost (%)
						Criterion	Quantity	
Other recurrent costs								
	Administration costs	Management						
		Administration services						
		Other ______						
	Training costs	in administration						
		in maintenance						
		in operations						
		Other ______						
	Promotion and education costs	in health education						
		Other ______						

Annex V

Time profiles of a beta distribution function

The time profile generated by a beta distribution function is given by the value of parameters α and β, which determine the shape of the cumulative distribution function $F(\tau;\alpha,\beta,\theta)$ and of its underlying density function:

$$f(\tau;\alpha,\beta,\theta) = \frac{dF(\tau;\alpha,\beta,\theta)}{d\tau}.$$

The density function expresses the instantaneous rate of change (speed) of the time trend $F(\tau;\alpha,\beta,\theta)$.

The following table summarizes the diversity of the time profiles that can be generated by a beta density function of the form:

$$f(\tau;\alpha,\beta,\theta) = \frac{\tau^{\alpha-1}(\theta-\tau)^{\beta-1}}{B(\alpha,\beta)\theta^{\alpha+\beta-1}}, \quad 0 \le \tau \le \theta,\ \alpha > 0,\ \beta > 0,$$

where $B(\alpha,\beta)$ denotes the beta function defined by the Euler integral of the first kind:

$$B(\alpha,\beta) = \int_0^1 x^{\alpha-1}(1-x)^{\beta-1}\,dx.$$

For a comprehensive technical presentation of the beta distribution, see Johnson *et al.* (1995), chapter 25.

The typical shapes of a beta density function

Parameters	$0 < \beta < 1$	$\beta = 1$	$1 < \beta < 2$	$\beta = 2$	$\beta > 2$
$0 < \alpha < 1$	**U-shaped** $\infty \downarrow$ MIN $\uparrow \infty$ with MIN at $\tau = \frac{(\alpha - 1)\theta}{\alpha + \beta - 2}$ Symmetrical about $\tau = \frac{\theta}{2}$ if $\alpha = \beta$	**Reverse J-shaped** $\infty \downarrow \frac{\alpha}{\theta}$	Reverse J-shaped $\infty \downarrow 0$		
$\alpha = 1$	**J-shaped** $\frac{\beta}{\theta} \uparrow \infty$	**Uniform-shaped** $\frac{1}{\theta} \rightarrow$	**Rotated reverse J-shaped** $\frac{\beta}{\theta} \downarrow 0$	**Linear-shaped** $\frac{2}{\theta} \downarrow 0$	**Reverse J-shaped** $\frac{\beta}{\theta} \downarrow 0$
$1 < \alpha < 2$	**J-shaped** $0 \uparrow \infty$	**Rotated J-shaped** $0 \uparrow \frac{\alpha}{\theta}$	**Bell-shaped** $0 \uparrow$ MAX $\downarrow 0$ with MAX at $\tau = \frac{(\alpha - 1)\theta}{\alpha + \beta - 2}$. Symmetrical about $\tau = \frac{\theta}{2}$ if $\alpha = \beta$. Points of inflexion at $\tau = \frac{\theta}{\alpha + \beta - 2}\left(\alpha - 1 \pm \sqrt{\frac{(\alpha - 1)(\beta - 1)}{\alpha + \beta - 3}}\right)$ provided these values are real and lie between 0 and θ. If they exist, they are equidistant from the mode.		
$\alpha = 2$		**Linear-shaped** $0 \uparrow \frac{2}{\theta}$			
$\alpha > 2$		**J-shaped** $0 \uparrow \frac{\alpha}{\theta}$			

From this table, we identify seven different time profiles that can be used to model the possible speed at which variables P_t, N_t and q_t could evolve during the life-cycle of a water supply system. The first two profiles are:

- *U-shaped*: speed drops from infinity to a minimum value, reached at time $\tau = (\alpha - 1)\theta/(\alpha + \beta - 2)$, followed by a rise up to infinity, reached at time $\tau = \theta$. This profile is generated when the parameter values are chosen so that $0 < \alpha < 1$ and $0 < \beta < 1$. If $\alpha = \beta$ the profile is symmetrical about $\tau = \theta/2$.
- *Bell-shaped*: speed increases from 0 to a maximum value, reached at time $\tau = (\alpha - 1)\theta/(\alpha + \beta - 2)$, followed by a decrease to 0, reached at time $\tau = \theta$. This profile is generated when the parameter values are chosen so that $\alpha > 1$ and $\beta > 1$. If $\alpha = \beta$ the profile is symmetrical about $\tau = \theta/2$.

Both the U-shaped and bell-shaped profiles have extreme (maximum or minimum) values within the time interval $[0;\theta]$. However, if $(\alpha - 1)(\beta - 1) \leq 0$, the beta density function does not have one of these profiles, but displays a monotone (increasing or decreasing) trend characterized by one of the following profiles:

- *J-shaped*: an increasing convex curve is obtained when the parameter values are chosen so that: $\alpha \geq 1$ and $0 < \beta < 1$. In particular, if $\alpha = 1$ the J-shaped curve increases from β/θ at $\tau = 0$, to infinity at $\tau = \theta$. If $\alpha > 1$, the starting value of this curve is 0.
- *Reverse J-shaped*: if the parameter values are chosen so that $0 < \alpha < 1$ and $\beta \geq 1$, a reverse J-shaped profile is obtained. This is a decreasing convex curve (the reverse of a J-shaped profile). If $\beta = 1$ the curve decrease takes place between infinity and α/θ. If $\beta > 1$ the final value of this curve is 0.
- The choice of a unit value for α or β leads to specific profiles. If $\alpha = \beta = 1$, we obtain a **uniform-shaped** profile, taking the constant value $1/\theta$ at any time $0 \leq \tau \leq \theta$.
- If $\beta = 1$ and $\alpha > 1$, the time profile is that of an increasing curve, from 0 to α/θ, but with different shapes depending on the value of α. In particular: if $\alpha = 2$ the curve is **linear-shaped**; if $\alpha > 2$ the curve is **J-shaped**; and if $1 < \alpha < 2$ the profile is that of a curve obtained by rotating the $\alpha > 2$ J-shaped profile about the $\alpha = 2$ linear profile, which we therefore call a **rotated J-shaped** profile.
- If $\alpha = 1$ and $\beta > 1$, the time profile is that of a decreasing curve, from β/θ to 0, but with different shapes depending on the value of β: **linear-shaped** if $\beta = 2$; **reverse J-shaped** if $\beta > 2$; and **rotated reverse J-shaped** if $1 < \beta < 2$.

From these typical speed time profiles $f(\tau;\alpha,\beta,\theta)$ we derive the following time trend profiles $F(\tau;\alpha,\beta,\theta)$:

- an **S-shaped** time trend generated by a **bell-shaped** speed profile;
- a **rotated S-shaped** time trend (about the segment joining the initial to the final value of the trend) generated by a **U-shaped** speed profile;
- a **linear-shaped** time trend generated by a **uniform-shaped** speed profile;
- a **J-shaped** time trend generated by a **J-shaped**, an increasing **linear-shaped** or a **rotated J-shaped** speed profile;
- a **rotated J-shaped** trend (about the segment joining the initial to the final value of the trend) generated by a **reverse J-shaped**, a decreasing **linear-shaped** or a **rotated reverse J-shaped** speed profile.

Annex VI

Costing case studies in Peru

VI.1 OBJECTIVES

A consultant in Peru was commissioned to test an early version of the costing method set out in this manual, including the excel calculation spreadsheet that we refer to as the Water Supply Costing Processor (Lampoglia, 2007). The consultant selected two projects to improve drinking-water supplies in rural areas of Peru as case studies.

- The first project is located in the mountain range of Cajamarca and is a system of water supply by gravity, without treatment and with household connections. In the terms used in this manual, the technology is classified as 'protected spring' and 'piped water into dwelling, plot or yard'. The service covers 115 beneficiaries.
- The second project is located in a forest and the system supplies water by gravity, with treatment and with household connections, to Guantanamo, San Martin. In the terms used in this manual, the technology is classified as 'slow sand filtration' and 'piped water into dwelling, plot or yard'. The service covers 193 beneficiaries.

The objectives of pilot-testing this manual and the Water Supply Costing Processor were:

- To collect and assess the costs of existing projects for improved water supply systems in low-income communities, using the costing method, the draft sets of questionnaires and the excel calculation spreadsheet developed by the Department of Econometrics of the University of Geneva. The scope of the pilot test was at national level, and the different types of water supply technologies used in the country were considered.
- To prepare an appraisal report of the user-friendliness of the questionnaires and the excel calculation spreadsheet in terms of applicability and practicality, indicating the level of data and information that could easily be obtained or generated. The pilot test also aimed to identify and report on:
  - existing costs not mentioned and not accounted for in the questionnaires or spreadsheet (e.g. hidden costs, subsidies, the transport costs of materials, labour contributed by users);
  - practical difficulties in the use of the manual and spreadsheet, such as limitations and adaptability;
  - the multiple constraints that affect the implementation of improved technologies, in particular the technical and socioeconomic constraints;
  - ways of improving the questionnaires and the excel calculation spreadsheet.

VI.2 METHOD

The pilot test followed the method set out in an early version of this manual. Thus, steps were taken to:

- document the importance of identifying and evaluating the investment in improved water supply technologies in the local setting of the project;
- define the type of the project;
- identify and gather data on the cost components of each activity of the project;
- define criteria and indicators for decision-makers to select the best option.

VI.3 COSTING INFORMATION

VI.3.1 Investment costs

As far as the availability of information goes, the costs of investment in infrastructure were the easiest to obtain. None of the systems evaluated required imported materials or equipment, and none of them used electrical energy.

The project budgets were generally calculated on the basis of unit costs for each component, considering the prices of materials, tools and labour. The costs of 'materials' and 'labour' were disaggregated by computing the respective percentages of the unit costs. The 'tools' component, which was generally small, was added to 'materials'.

It was difficult to classify labour as 'qualified' or 'non-qualified', as required in the costing questionnaires. For this reason, labour was either put in a single category or divided between 'qualified' and 'non-qualified' using a simplified sharing rule.

The costs of promotion and education were grouped into 'promotion and training in administration, operation and maintenance' and 'hygiene education'. The associated activities refer to the implementation of the training and hygiene education programme targeting the beneficiary community. It is important to note that the projects are part of a long-running programme in the region and were implemented by a nongovernmental organization. Thus, other relevant investment costs, such as the cost of preparing educational materials or the administrative costs of the programme, are not included in the budgets of the projects.

Other investment costs, such as engineering, supervision and administration of the project, were computed on the basis of a percentage of the direct cost or according to the value indicated in the technical file of the project.

VI.3.2 Recurrent costs

The proposed method is to evaluate independently the costs of operation and maintenance, for the categories of materials and labour.

In small rural systems, however, it is difficult to make this distinction. The work of operation consists of opening or closing valves to chlorinate the water. Generally, the person in charge of these tasks also does the maintenance work. Quantifying the hours for these separate activities is difficult. Therefore, in the pilot tests, it was decided to categorize all the labour as 'maintenance'.

Another issue that arose in the pilot tests was that it was not possible to separate the training costs into administration, operation and maintenance components because the institution responsible for the implementation of the project grouped these values as 'training for administration, operation and maintenance'. As no information was available on these individual costs, we adopted the values used in a project consultancy carried out for the World Bank by Lampoglia (2004). The same approach was used to assess the costs of hygiene education.

The administrative costs were included in the classification 'Other recurrent costs'.

VI.4 TECHNICAL AND SOCIOECONOMIC INFORMATION

Technical statistical data were obtained from the National Institute of Statistic and Computer Science (INEI) in Peru.

The socioeconomic data for each case study was obtained from a household survey, complemented by local population statistics from the National Census of Housing carried out by the INEI (2005).

The available information on the demographic structure did not match that requested by the questionnaire. With regard to income, the information was grouped by range of income rather than by quintiles of population, as indicated in the final questionnaire.

VI.5 CASE STUDY 1: RURAL POTABLE WATER SYSTEM IN BELLAVISTA, CAJAMARCA

VI.5.1 General information about the project

- *Responsible organization for implementation of the project:* CARE Peru
- *Financing:* Swiss Agency for Development and Cooperation (SDC)
- *Components:* infrastructure, training in administration, operation and maintenance, and hygiene education
- *Location:* Bellavista Alta, District of Miguel Iglesias, Province of Celendín, Department of Cajamarca
- *Starting date:* 2004

Bellavista is a rural centre in the Andes Mountains, in the District of Miguel Iglesias, Province of Celendín, Department of Cajamarca. The total population is estimated as 415 inhabitants, living in 83 houses. The economic activity of the population of Bellavista, as for the population of the entire region, is agriculture and livestock.

The project covers the supply of drinking-water to 23 houses in Bellavista Alta. It has been implemented by the nongovernmental organization CARE, with the financing from SDC, as part of the PROPILAS programme to improve the management and sustainability of water and sanitation in the region.

Like all the rural systems in Peru, the management, operation and maintenance of the water system is executed under the responsibility of a local administrator from the sanitation service, Junta Administradora de Servicios de Saneamiento (JASS), with participation of members of the community. The project also includes a training component by JASS.

The beneficiaries of the project contributed both in cash and in kind (by providing unskilled labour and local materials, and by transporting materials to the area).

The evaluated costs include the investment in infrastructure, training for system administration, operation and maintenance, and hygiene education.

VI.5.2 Description of the system

The water supply system is a **protected spring and piped water into dwelling, plot or yard**. It consists of a water point at the spring, a line of water conveyance a rectangular reservoir (volume = 3 m^3) and valves, 5.81 metres of water distribution network, and 23 household taps. There are also ten chambers positioned to act as pressure breaks, two control boxes with valves, four purge valves with their respective boxes, an air valve, and two aerobe passages.

VI.5.3 Lifetime of the system

The expected lifetime of the system is 15 years.

VI.5.4 Technical and socioeconomic questionnaires

TECHNICAL QUESTIONNAIRE

Date: December 2005

GENERAL INFORMATION

1. Country: **Peru**
2. Region: **Department of Cajamarca, Province of Celendín, District of Miguel Iglesias**
3. Location of project (town or village): **Bellavista – Caserío**
4. Source of data: **INEI Instituto Nacional de Estadística e Informática, Peru**
5. Number of inhabitants: **415**
6. Number of dwellings/houses: **83**
7. Number of families: **83**
8. Number of persons per family: **5**
9. Density (in inhabitants per km^2): **20.4**

COMMUNITY CHARACTERISTICS

10. Description of the community: **Rural Community, high poverty. Subsistence life, only 3.6% of population employed.**
11. Type of dwellings/houses: **Rural huts with walls of local materials (78%). 61% with straw roof.**
12. Schools, postal office, public services: **Apart from the primary school of Bellavista, all the others public services are provided for the whole District of Miguel Iglesias**
13. Equipment of the community: **none**
14. Distance to the nearest community: **2,0 km**
15. Quality of communications (highways, roads, etc.):
    - ☐ Very good
    - ☐ Good
    - ☐ Average
    - ☒ Poor
    - ☐ Very poor
16. Type of transport (indicate all options):
    - ☒ Automobile
    - ☒ Bus
    - ☐ Train
    - ☒ Others (specify): trucks, motorcycle, van.
17. Current water supply system:
    - ☐ Surface water (river, dam, lake, pond, stream, canal, irrigation channels)
    - ☐ Tanker-truck
    - ☐ Bottled water
    - ☐ Cart with small tank/drum
    - ☒ Unprotected spring
    - ☒ Unprotected dug well
    - ☐ Rainwater collection
    - ☐ Protected spring
    - ☐ Protected dug well
    - ☐ Tube well or borehole
    - ☐ Public tap or standpipe

- ☐ Piped water into dwelling, plot or yard
- ☐ Others (specify): ________________

18. Type of source for current water supply:
    - ☐ Surface water
    - ☒ Groundwater
    - ☐ Rainwater
    - ☐ Others (specify): ________________
19. Current sanitation system:
    - ☒ Open defecation
    - ☐ Hanging latrine
    - ☐ Bucket latrine
    - ☐ Pit latrine without slab or platform
    - ☐ Flush or pour-flush to elsewhere
    - ☐ Shared sanitation
    - ☐ Composting toilet
    - ☐ Pit latrine with slab
    - ☐ Ventilated improved pit latrine
    - ☐ Flush or pour-flush toilet/latrine to:
    - ☐ Piped sewer system
    - ☐ Septic tank
    - ☐ Pit latrine
    - ☐ Others (specify): ________________
20. Company responsible for power (energy) : none
21. Power installed: none
22. Power provided (on average, per 24-hour period): none
    From _______ : _______, to _______ : _______
23. Map of the region (specify): District of Miguel Iglesias

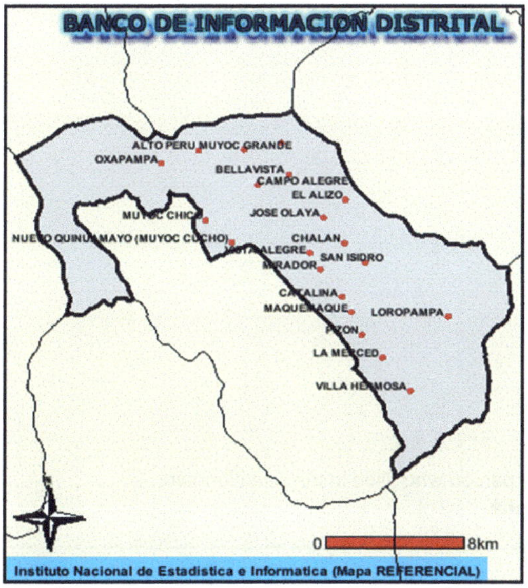

Source: INEI (2005)

HYDROLOGY AND CLIMATE

24. Geographical location:
    - ☐ Coastal
    - ☐ Continental
    - ☐ Forest
    - ☒ Mountain
    - ☐ Others (specify): ______________

25. Average temperatures (in degrees Celsius):

☒ Winter **5°**	☐ Dry season
☐ Spring	☐ Intermediate season
☒ Summer **21°**	☐ Rainy season
☐ Fall	

Minimum and maximum temperatures (in degrees Celsius):

☐ Winter	☐ Dry season
☐ Spring	☐ Intermediate season
☐ Summer	☐ Rainy season
☐ Fall	

26. Rainfall (average) : **625.9 mm/year**
27. Rainy months:
    - ☒ January
    - ☒ February
    - ☒ March
    - ☐ April
    - ☐ May
    - ☐ June
    - ☐ July
    - ☐ August
    - ☐ September
    - ☐ October
    - ☐ November
    - ☒ December
28. Dry months:
    - ☐ January
    - ☐ February
    - ☐ March
    - ☒ April
    - ☒ May
    - ☒ June
    - ☒ July
    - ☒ August
    - ☒ September
    - ☐ October
    - ☐ November
    - ☐ December

Name and signature of the person who filled in the questionnaire:
Engineer Teresa Lampoglia

The technical questionnaire for the rural potable water system of Bellavista-Cajamarca in Peru shows that, based on 1993 Census data, the population of this locality in December 2005 was estimated to be 415 inhabitants, comprising 83 families.

The community is rural with a high poverty level. The rural production is for subsistence and the level of local employment is only 3.6%. The houses are rural huts with walls constructed using local materials (78%). Many of the huts have straw roofs (61%).

Apart from the primary school of Bellavista, all other public services are provided for the whole District of Miguel Iglesias. The community does not have suitable equipment and the communication routes are few and are badly maintained. The Access is by car, bus, truck, and so on.

The present water supply system is an unprotected well/spring and the water supply source is therefore groundwater. Sanitation and electricity systems are unavailable.

The geographical situation of the community is mountainous with an average temperature in winter of 5° Celsius and in summer of 21° Celsius. The annual rainfall is 626 mm, mainly in the months of December to March.

SOCIOECONOMIC QUESTIONNAIRE

Date: December 2005

GENERAL INFORMATION

1. Country: **Peru**
2. Region: **Department of Cajamarca, Province of Celendín, District of Miguel Iglesias**
3. Location of project (town or village): **Bellavista – Caserío**
4. Source of data: **Domiciliary survey and Censo Nacional de Vivienda del INEI (2005), Peru**
5. Number of inhabitants: **415**
6. Number of dwellings/houses: **83**
7. Number of families: **83**
8. Number of persons per family: **5**
9. Density (in inhabitants per km^2): **20.4**

COMMUNITY CHARACTERISTICS

10. Type of community:
    - ☐ Urban
    - ☐ Suburban
    - ☒ Rural
11. Proportion of men and women:
    - Men **47%**
    - Women **53%**
12. Demographic structure of the population:
    - <5 years **16%**
    - 5–14 years **28%**
    - 15–64 years **51%**
    - >65 years **5%**
13. Proportion of population according to highest educational level (District data):
    - ☒ Complete university education **0.8%**
    - ☐ Incomplete university education
    - ☒ Complete secondary education **0.4%**
    - ☐ Incomplete secondary education
    - ☒ Complete primary education **67%**
    - ☐ Incomplete primary education
    - ☒ No education **32%**

14. Level of income according to population percentile: **N.A.**
    1% to 20% ____________ USD$ or local currency
    20% to 39% ____________ USD$ or local currency
    40% to 59% ____________ USD$ or local currency
    60% to 79% ____________ USD$ or local currency
    80% to 100% ____________ USD$ or local currency
    ☒ Subsistence income
15. Type and level of employment:
    ☐ Formal ______%
    ☒ Informal **100%**
    ☐ Seasonal ______%
    ☐ Full-time ______%
16. Proportion of employment by production sector:
    ☒ Agriculture, cattle and livestock **90%**
    ☐ Commerce and industry ______%
    ☐ Transport and telecommunication ______%
    ☐ Mining ______%
    ☐ Forestry ______%
    ☐ Tourism ______%
    ☐ Housing and construction ______%
    ☐ Services ______%
    ☒ Others (specify):
    Unskilled work in manufacturing, mining and construction **8%**
    Unskilled work in services and commerce **2%**

SOCIAL ORGANIZATIONS/ACTIVITIES

17. Local organizations:

	Yes	No
Neighbourhood associations		**X**
Parents and family centres	**X**	
Sports clubs	**X**	
Gender groups	**X**	

18. Names of social organizations:
    Committee of water and sanitation
    Glass of milk
19. Attitude of the community to the sanitation: **Support activities on the basis of social promotion which satisfies its culture and ideology.**
20. Possible contributions of the community to the water supply project:
    ☒ Financial resources ______% or **PEN 2 381**
    ☒ Materials ______% or **PEN 7 427**
    ☒ Labour **16%** or **PEN 15 971 or 12 777 hours**
    ☐ Equipment ______% or local currency or USD
    ☒ Others (specify): **Rural carriage,** **PEN 2 270**

Name and signature of the person who filled in the questionnaire:
Engineer Teresa Lampoglia

According to the socioeconomic questionnaire, 44% of the population is less than 14 years old and 67% of the population have primary education, while 32% have no education.

The income is only at subsistence level, either from agricultural or informal work.

Some social organizations are represented by local neighbours and mothers centres. The community has a positive attitude to sanitation promotion in terms of social culture and ideology.

The community contributed to the project in terms of financial resources, materials, transport and labour.

VI.5.5 Economic evaluation

The economic evaluation of this project was carried out on the assumption that all 23 households that were part of the project design (115 inhabitants in total) benefited from the services of the water supply system since its date of completion, conventionally defined as the beginning of the lifetime period. Therefore, the opportunity cost of spare capacity is nil.

Data provided in the 'Project design' and 'Social costing' spreadsheets of the Water Supply Costing Processor (WSCP) to compute the full and unit cost indicators of this water supply project are displayed below. Note that the investment cost components, which were input in the costing questionnaire spreadsheets at their historical value of January 2004, were converted to the Peruvian Nuevo sol (PEN) value at the reference date of 1 December 2006, conventionally chosen as the starting date of the use of the water supply system, using the monthly series of the consumer price index.

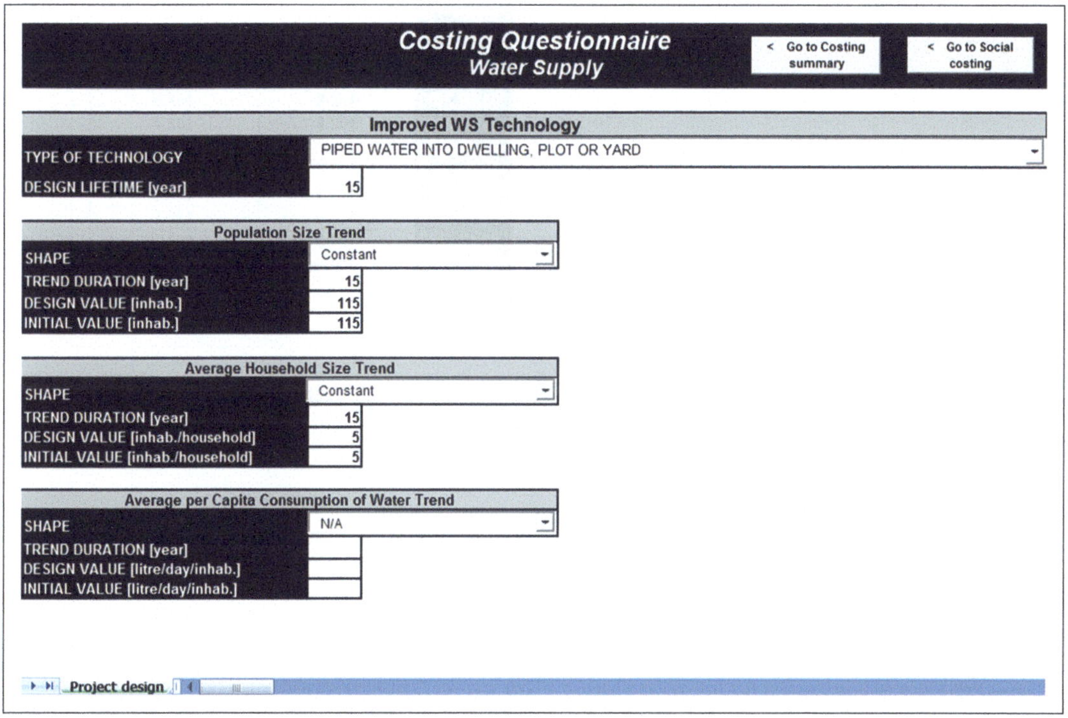

As shown by the 'Costing summary' spreadsheet displayed below, the full cost present value of the project is estimated at 158 268 [PEN]/[life-cycle] (in PEN of 1 December 2006, noting that the rate of exchange in December 2006 was 1 PEN to 0.3125 US$). Of this full cost, 65% was attributable to investment costs, 23% to maintenance costs, 0.1% to operation costs and 12% to administration costs. This full cost present value can be converted into an annual equivalent cost of 19 828 [PEN]/[year], representing the constant annuity to be paid during the project life-cycle of 15 years to refund the full cost present value of 158 268 [PEN]/[life-cycle].

Costing Questionnaire
Water Supply

< Go to Project design

Valuation parameters	
Date of actualization [mm.yyyy]	12.2006
Local currency	Soles
Shadow factor for unskilled labour wage	0.41
Shadow factor for foreign rate of exchange	
Real annual social discount rate [%]	11

Consumer Price Index		
Date	*Base (=100) 12/2001*	*Base (=100) 12/2006*
01.12.2001	100.0	110.5
01.01.2002	99.5	111.1
01.02.2002	99.4	111.1
01.03.2002	100.0	110.5
01.04.2002	100.7	109.7
01.05.2002	100.8	109.6
01.06.2002	100.6	109.8
01.07.2002	100.7	109.8
01.08.2002	100.8	109.7
01.09.2002	101.2	109.2
01.10.2002	102.0	108.4
01.11.2002	101.6	108.8
01.12.2002	101.5	108.9
01.01.2003	101.8	108.6
01.02.2003	102.2	108.1
01.03.2003	103.4	106.9
01.04.2003	103.3	107.0
01.05.2003	103.3	107.0
01.06.2003	102.8	107.5
01.07.2003	102.6	107.7
01.08.2003	102.7	107.6
01.09.2003	103.2	107.1
01.10.2003	103.3	107.0
01.11.2003	103.5	106.8
01.12.2003	104.0	106.2
01.01.2004	104.6	105.7
01.02.2004	105.7	104.5

Social costing

Unit costs can be derived by dividing the full cost present value of the project by the present value of its expected life-cycle production, estimated as 918 [inhabitant-year]/[life-cycle] or 184 [household-year]/[life-cycle]. This leads to an average incremental cost of 172 [PEN]/[inhabitant-year] or 862 [PEN]/[household-year], respectively.

VI.6 CASE STUDY 2: RURAL POTABLE WATER SYSTEM IN GUANTÁNAMO, SAN MARTÍN

VI.6.1 General information about the project

- *Responsible organization for implementation of the project:* nongovernmental organization PRISMA
- *Financing:* Swiss Agency for Development Cooperation (SDC)
- *Components:* infrastructure, training in administration, operation and maintenance, and hygiene education
- *Location:* Guantánamo, District Nuevo Progreso, Province of Tocache, Department of San Martín
- *Starting date:* 2003

Guantánamo is a rural population centre located in the Peruvian low forest, on the right bank of the Huallaga River. Guantánamo is located in the District Nuevo Progreso, Province of Tocache, Department of San Martín.

Costing Questionnaire
Water Supply

< Go to Project design

Selected improved technology: *PIPED WATER INTO DWELLING, PLOT OR YARD*

Currency and day of actualization: *Soles of 01.12.2006*

TOTAL COST TYPOLOGY	FULL COST			AVERAGE INCREMENTAL COST					
	PRESENT VALUE	ANNUAL EQUIVALENT	[%]	[Soles]/[inhabitant-year]		[Soles]/[household-year]		[Soles]/[litre/day-year]	
	[Soles]/[life-cycle]	[Soles]/[year]		Expected use of design capacity	Full use of design capacity	Expected use of design capacity	Full use of design capacity	Expected use of design capacity	Full use of design capacity
TOTAL INVESTMENT COSTS	**103'464.84**	**12'962.49**	**65.37**	**112.72**	**112.72**	**563.59**	**563.59**	**N/A**	**N/A**
Local materials	33'601.18	4'209.69	21.23	36.61	36.61	183.03	183.03	N/A	N/A
Imported materials	-	-	0.00	-	-	-	-	N/A	N/A
Local equipments	-	-	0.00	-	-	-	-	N/A	N/A
Imported equipments	-	-	0.00	-	-	-	-	N/A	N/A
Labour	27'848.32	3'488.95	17.60	30.34	30.34	151.69	151.69	N/A	N/A
Other investment costs	15'685.77	1'965.18	9.91	17.09	17.09	85.44	85.44	N/A	N/A
Incidental investment costs	26'329.57	3'298.67	16.64	28.68	28.68	143.42	143.42	N/A	N/A
TOTAL MAINTENANCE COSTS	**35'918.39**	**4'500.00**	**22.69**	**39.13**	**39.13**	**195.65**	**195.65**	**N/A**	**N/A**
Local materials	13'569.17	1'700.00	8.57	14.78	14.78	73.91	73.91	N/A	N/A
Imported materials	-	-	0.00	-	-	-	-	N/A	N/A
Local equipments	-	-	0.00	-	-	-	-	N/A	N/A
Imported equipments	-	-	0.00	-	-	-	-	N/A	N/A
Labour	22'349.22	2'800.00	14.12	24.35	24.35	121.74	121.74	N/A	N/A
TOTAL OPERATION COSTS	**207.53**	**26.00**	**0.13**	**0.23**	**0.23**	**1.13**	**1.13**	**N/A**	**N/A**
Local materials	207.53	26.00	0.13	0.23	0.23	1.13	1.13	N/A	N/A
Imported materials	-	-	0.00	-	-	-	-	N/A	N/A
Local power services	-	-	0.00	-	-	-	-	N/A	N/A
Imported power services	-	-	0.00	-	-	-	-	N/A	N/A
Labour	-	-	0.00	-	-	-	-	N/A	N/A
TOTAL OTHER RECURRENT COSTS	**18'677.56**	**2'340.00**	**11.80**	**20.35**	**20.35**	**101.74**	**101.74**	**N/A**	**N/A**
Administration	18'677.56	2'340.00	11.80	20.35	20.35	101.74	101.74	N/A	N/A
Training	-	-	0.00	-	-	-	-	N/A	N/A
Promotion & education	-	-	0.00	-	-	-	-	N/A	N/A
TOTAL COST	**158'268.33**	**19'828.49**	**100**	**172.42**	**172.42**	**862.11**	**862.11**	**N/A**	**N/A**

	[inhabitant-year]/[life-cycle]	[household-year]/[life-cycle]	[litre/day-year]/[life-cycle]
PRESENT VALUE OF EXPECTED LIFE-CYCLE PRODUCTION	918	184	N/A

	[inhabitant-year]	[household-year]	[litre/day-year]
DESIGN CAPACITY	115	23	N/A

Costing summary

The main land use is agriculture, with coffee, yucca, banana and corn. Some land is used for raising cattle and other animals.

The project involves supplying potable water to the population of Guantánamo, by piping the water to 50 buildings, including 46 houses, two churches, a communal house and an education centre.

The houses are located in a slightly dispersed form. The main construction material is wood and the roofs are made of palm. Houses are typically single-storey, with a roof.

Like all the rural water systems in Peru, the management, operation and maintenance of the system is performed under the responsibility of a local administrator from the sanitation service, Junta Administradora de Servicios de Saneamiento (JASS), with participation of members of the community. The project also includes a training component by JASS.

The beneficiaries of the project contributed unskilled labour, local materials and cash. The cash contribution is equivalent to the cost of their household tap. The municipality also participated in the co-financing of the project (25% of the direct cost of the works), and contributed to the supervision of the project.

The evaluated costs include investment in infrastructure, training in administration, operation and maintenance, and hygiene education.

VI.6.2 Description of the system

The system **is slow sand filtration and piped water into the dwelling, plot or yard**. The source of water supply is the Alto Chiquito Río gorge. Water is taken in through a collector and transported first to sedimentation and then to a slow sand filter and reservoir with a capacity of 10 m^3. From the reservoir, a line of water conveyance of 1.5 inches in diameter transports the water into the distribution network and to the household taps.

VI.6.3 Lifetime of the system

The expected lifetime of the system is 20 years.

VI.6.4 Technical and socioeconomic questionnaires

TECHNICAL QUESTIONNAIRE

Date: December 2005

GENERAL INFORMATION

1. Country: **Peru**
2. Region: **Department of San Martín, Province of Tocache, District Nuevo Progreso**
3. Location (town or village): **Guantánamo – Caserío**
4. **Source of data: INEI Instituto Nacional de Estadística e Informática, Peru**
5. Number of inhabitants: **193**
6. Number of dwellings/houses: **58**
7. Number of families: **58**
8. Number of persons per family: **3.33**
9. Density (in inhabitants per km^2): **11.1**

COMMUNITY CHARACTERISTICS

10. Description of the community: **Rural community with high poverty level. Only subsistence life income level. 100% of community do not have safe water and 90% do not have sanitary services. The town altitude above sea level: 490 metres.**
11. Type of dwellings/houses: **Rural huts with wood walls (70%) and 77% with calamine roofs. 87% are owners and 97% are independent house. The 97% do not have electricity.**
12. Schools, postal office, public services: **Apart from the primary school of Guantánamo, all the others public services are provided for the whole District Nuevo Progreso.**
13. Equipment of the community: **none**
14. Distance to the nearest community: **5.0 km.**
15. Quality of communications (highways, roads, etc.):
    - ☐ Very good
    - ☐ Good
    - ☐ Average
    - ☒ Poor
    - ☐ Very poor
16. Type of transport (indicate all options):
    - ☒ Automobile
    - ☒ Bus
    - ☐ Train
    - ☒ Others (specify): truck, motorbike
17. Current water supply system:
    - ☒ Surface water (river, dam, lake, pond, stream, canal, irrigation channels)
    - ☐ Tanker-truck
    - ☐ Bottled water
    - ☐ Cart with small tank/drum
    - ☐ Unprotected spring
    - ☐ Unprotected dug well
    - ☐ Rainwater collection
    - ☐ Protected spring
    - ☐ Protected dug well
    - ☐ Tube well or borehole
    - ☐ Public tap or standpipe
    - ☐ Piped water into dwelling, plot or yard
    - ☐ Others (specify) ____________________
18. Type of source of water supply:
    - ☒ Surface water
    - ☐ Groundwater
    - ☐ Rainwater
    - ☐ Others (specify): ____________________
19. Current sanitation system:
    - ☒ Open defecation – 90%
    - ☐ Hanging latrine
    - ☐ Bucket latrine
    - ☐ Pit latrine without slab or platform
    - ☐ Flush pr pour-flush to elsewhere
    - ☐ Shared sanitation
    - ☐ Composting toilet
    - ☒ Pit latrine with slab – 10%
    - ☐ Ventilated improved pit latrine
    - ☐ Flush or pour-flush to:
    - ☐ Piped sewer system

- ☐ Septic tank
- ☐ Pit latrine
- ☐ Others (specify): ____________________

20. Company responsible for power (energy): **none**
21. Power installed: **none**
22. Power provided (on average, per 24–hour period): **none**
23. Map of the region (specify): **District Nuevo Progreso**

Source: INEI (2005)

HYDROLOGY AND CLIMATE

24. Geographical location:
    - ☐ Coastal
    - ☐ Continental
    - ☒ Forest
    - ☐ Mountain
    - ☐ Others (specify): ____________________
25. Average temperatures (in degrees Celsius):

☒ Winter **19°**	☐ Dry season
☐ Spring	☐ Intermediate season
☒ Summer **35°**	☐ Rainy season
☐ Fall	

Minimum and maximum temperatures (in degrees Celsius):

- ☐ Winter
- ☐ Spring
- ☐ Summer
- ☐ Fall
- ☐ Dry season
- ☐ Intermediate season
- ☐ Rainy season

26. Rainfall (average): **1'149.7 mm/year**
27. Rainy months:
    - ☒ January
    - ☒ February
    - ☒ March
    - ☐ April
    - ☐ May
    - ☐ June
    - ☐ July
    - ☐ August
    - ☐ September
    - ☒ October
    - ☒ November
    - ☒ December
28. Dry months:
    - ☐ January
    - ☐ February
    - ☐ March
    - ☒ April
    - ☒ May
    - ☒ June
    - ☒ July
    - ☒ August
    - ☒ September
    - ☐ October
    - ☐ November
    - ☐ December

Name and signature of the person who filled in the questionnaire:
Engineer Teresa Lampoglia

The technical questionnaire for the rural water system in Guantánamo-San Martín in Peru shows that, based on 1993 census data, the population of this locality in December 2005 was estimated to be 193 inhabitants, comprising 58 families.

The community is rural, with a high degree of poverty. Production is only for subsistence (100%). In the community, 90% do not have sanitary services and 97% do not have electricity.

The houses are rural huts with walls of wood (70%), and 77% have calamine roofs. Almost all (97%) of the houses are detached, and 87% are owner-occupied.

Apart from the primary school of Guantánamo, all the others public services are provided for the whole District Nuevo Progreso. Access to the locality is via highways, by car or light truck. The distance from the next locality, Tocache, is approximately 90 km on good roads, and the journey takes approximately two and a half hours.

The type of land where the water supply system will be located is of clay with a variable slope. Vegetation is abundant.

Potable water is provided by spring or river. Only 10% of the community has a sanitation system (a pit latrine).

The climate is warm and humid, with humidity similar to that of the surrounding low-lying forest area, and rainfall occurs throughout the year.

The rainiest season is from January to March, and summer is from June to September.

SOCIOECONOMIC QUESTIONNAIRE

Date: December 2005

GENERAL INFORMATION

1. Country: **Peru**
2. Region: **Department of San Martín, Province of Tocache, District Nuevo Progreso.**
3. Location (town or village): **Guantánamo – Caserío.**
4. Source of data: **Domiciliary survey and Censo Nacional de Vivienda del INEI (2005), Peru.**
5. Number of inhabitants: **193**
6. Number of dwellings/houses: **58**
7. Number of families: **58**
8. Number of persons per family: **3.33**
9. Density (in inhabitants per km^2): **11.1**

COMMUNITY CHARACTERISTICS

10. Type of community:
    - ☐ Urban
    - ☐ Suburban
    - ☒ Rural
11. Proportion of men and women:
    Men **61%**
    Women **39%**
12. Demographic structure of the population:
    <5 years **20%**
    5–14 years **31%**
    15–64 years **47%**
    >65 years **2%**
13. Proportion of population according to highest educational level (**District data**):
    - ☒ Complete university education **1%**
    - ☐ Incomplete university education
    - ☒ Complete secondary education **13%**
    - ☐ Incomplete secondary Education
    - ☒ Complete primary education **71%**
    - ☐ Incomplete primary education
    - ☒ No education **15%**
14. Level of income according to population percentile: **N.A.**
    1% to 20% ____________ USD$ or local currency
    20% to 39% ____________ USD$ or local currency
    40% to 59% ____________ USD$ or local currency
    60% to 79% ____________ USD$ or local currency
    80% to 100% ____________ USD$ or local currency
    - ☒ Subsistence income
15. Type and level of employment:
    - ☐ Formal ______%
    - ☒ Informal **100%**

- ☐ Seasonal ______%
- ☐ Full-time ______%

16. Proportion of employment by production sector:
    - ☒ Agriculture, cattle and livestock **71%**
    - ☐ Commerce and industry ______%
    - ☐ Transport and telecommunication ______%
    - ☐ Mining ______%
    - ☐ Forest ______%
    - ☐ Tourism ______%
    - ☐ Housing and construction ______%
    - ☐ Services ______%
    - ☒ Others (specify):
      Unskilled labour in manufacturing, mining and construction **7%**
      Unskilled labour in commerce **11%**

SOCIAL ORGANIZATIONS/ACTIVITIES

17. Local organizations:

	Yes	No
Neighbourhood associations		**X**
Parents and family centres	**X**	
Sport clubs	**X**	
Gender groups		**X**

18. Names of social organization:
    Committee of water and sanitation
    Glass of milk
19. Attitude of the community to the sanitation: **Support activities on the basis of a social promotion which satisfies the culture and ideology**
20. Possible contributions of the community to the water supply project:
    - ☐ Financial resources ______ % or ______ local currency
    - ☐ Materials ______ % or ______ local currency
    - ☒ Labour **16%** or ______ local currency or hours
    - ☐ Equipment ______% or ______ local currency or USD
    - ☒ Others (specify): **Communal work by unskilled labour and investment in home sanitary services**

Name and signature of the person who filled in the questionnaire:
Engineer Teresa Lampoglia

According to the socioeconomic questionnaire, 51% of the population is less than 14 years old and 71% of the population have primary education, while 15% have no education.

Income is at subsistence level, and the source of income is either agricultural or informal work. Some social organizations are represented by local neighbours and mothers centres. The community has a positive attitude to sanitation promotion in terms of social culture and ideology.

The community contributed to the project in terms of unskilled labour, and they invested in home sanitary services.

Economic evaluation

The water supply project of Guantánamo-San Martín was designed to provide drinking-water to a population of 50 families. The initial population benefiting from this project was estimated at 300 inhabitants (6 inhabitants per family) but the infrastructure was laid out to supply water to a design population of 408 inhabitants reached after 7 full years of growth at an average annual rate of 4.5% (historical growth rate of the decade 1993–2003). In order to model a smooth trend in the growth of the population of beneficiaries over the project lifetime of 20 years, a SYMMETRICAL S-SHAPED ($\alpha = \beta = 2$) time trend profile was selected.

The costing of this project has to take into account the time-varying level of services provided by the project. This is done by computing the life-cycle production as the present value of services provided to the growing number of inhabitants during the project life-time.

Data provided in the 'Project design' and 'Social costing' spreadsheets of the Water Supply Costing Processor (WSCP), used to compute the full and unit cost indicators of this water supply project, are displayed below. Note that the investment costs input in the costing questionnaire spreadsheets at the historical values of January 2003 were converted, using the monthly series of the consumer price index, to the PEN value at the reference date of 1 December 2006, conventionally chosen as the starting date of the use of the water supply system.

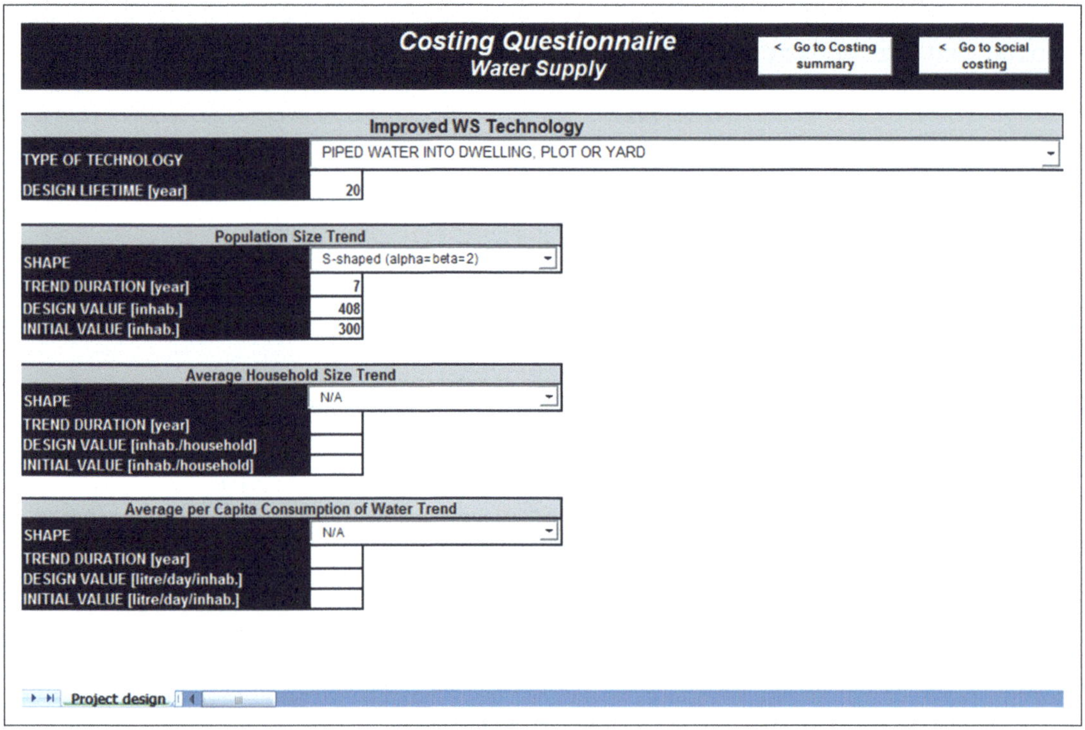

Costing Questionnaire
Water Supply

< Go to Costing summary | < Go to Social costing

Improved WS Technology	
TYPE OF TECHNOLOGY	PIPED WATER INTO DWELLING, PLOT OR YARD
DESIGN LIFETIME [year]	20

Population Size Trend	
SHAPE	S-shaped (alpha=beta=2)
TREND DURATION [year]	7
DESIGN VALUE [inhab.]	408
INITIAL VALUE [inhab.]	300

Average Household Size Trend	
SHAPE	N/A
TREND DURATION [year]	
DESIGN VALUE [inhab./household]	
INITIAL VALUE [inhab./household]	

Average per Capita Consumption of Water Trend	
SHAPE	N/A
TREND DURATION [year]	
DESIGN VALUE [litre/day/inhab.]	
INITIAL VALUE [litre/day/inhab.]	

Project design

As shown in the 'Costing summary' spreadsheet displayed below, the full cost present value of the Guantánamo-San Martin water supply project is estimated (in PEN of 1 December 2006) as 251 425 [PEN]/[life-cycle] (the rate of exchange in December 2006 was 1 PEN to 0.3125 US$). Of this full cost, 75% was attributable to investment costs, 12% to maintenance costs, 5% to operation costs and 8% to administrative costs. This full cost is converted into an annual equivalent cost of 28 444 [PEN]/[year], representing the constant annuity to be paid during the project life-cycle of 20 years to refund the full cost present value of 251 425 [PEN]/[life-cycle].

Unit costs can be derived by dividing the full cost present value of the project by the present value of its life-cycle production estimated (in PEN of 1 December 2006) as 3 244 [inhabitant-year]/[life-cycle]. This leads to an average incremental cost of 77.54 [PEN]/[inhabitant–year].

Costing Questionnaire
Water Supply

< Go to Project design

Valuation parameters	
Date of actualization [mm.yyyy]	12.2006
Local currency	PEN
Shadow factor for unskilled labour wage	0.49
Shadow factor for foreign rate of exchange	
Real annual social discount rate [%]	11

Consumer Price Index		
Date	*Base (=100) 12/2001*	*Base (=100) 12/2006*
01.12.2001	100.0	110.5
01.01.2002	99.5	111.1
01.02.2002	99.4	111.1
01.03.2002	100.0	110.5
01.04.2002	100.7	109.7
01.05.2002	100.8	109.6
01.06.2002	100.6	109.8
01.07.2002	100.7	109.8
01.08.2002	100.8	109.7
01.09.2002	101.2	109.2
01.10.2002	102.0	108.4
01.11.2002	101.6	108.8
01.12.2002	101.5	108.9
01.01.2003	101.8	108.6
01.02.2003	102.2	108.1
01.03.2003	103.4	106.9
01.04.2003	103.3	107.0
01.05.2003	103.3	107.0
01.06.2003	102.8	107.5
01.07.2003	102.6	107.7
01.08.2003	102.7	107.6
01.09.2003	103.2	107.1
01.10.2003	103.3	107.0
01.11.2003	103.5	106.8
01.12.2003	104.0	106.2
01.01.2004	104.6	105.7
01.02.2004	105.7	104.5

Social costing

To assess the opportunity cost of spare capacity, the full cost present value and the present value of the life-cycle production are recomputed by assuming a full use of the design capacity of the project. By dividing these two figures, an average incremental cost at full use of design capacity of 69.7 [PEN]/[inhabitant-year] is obtained. Compared to the above value of the average incremental cost of 77.54 [PEN]/[inhabitant–year], computed at the expected use of the design capacity, the opportunity cost of spare capacity appears to be of 7.8 [PEN]/[inhabitant–year].

Costing Questionnaire
Water Supply

< Go to Project design

Selected improved technology: *PIPED WATER INTO DWELLING, PLOT OR YARD*
Currency and day of actualization: *PEN of 01.12.2006*

TOTAL COST TYPOLOGY	FULL COST			AVERAGE INCREMENTAL COST					
	PRESENT VALUE	ANNUAL EQUIVALENT	[%]	[PEN]/[inhabitant-year]		[PEN]/[household-year]		[PEN]/[litre/day-year]	
	[PEN]/[life-cycle]	[PEN]/[year]		Expected use of design capacity	Full use of design capacity	Expected use of design capacity	Full use of design capacity	Expected use of design capacity	Full use of design capacity
TOTAL INVESTMENT COSTS	**187'039.60**	**21'160.02**	**74.39**	**57.65**	**51.86**	**N/A**	**N/A**	**N/A**	**N/A**
Local materials	96'281.09	10'892.40	38.29	29.68	26.70	N/A	N/A	N/A	N/A
Imported materials	-	-	0.00	-	-	N/A	N/A	N/A	N/A
Local equipments	113.17	12.80	0.05	0.03	0.03	N/A	N/A	N/A	N/A
Imported equipments	-	-	0.00	-	-	N/A	N/A	N/A	N/A
Labour	45'495.67	5'146.98	18.10	14.02	12.62	N/A	N/A	N/A	N/A
Other investment costs	45'149.67	5'107.84	17.96	13.92	12.52	N/A	N/A	N/A	N/A
Incidental investment costs	-	-	0.00	-	-	N/A	N/A	N/A	N/A
TOTAL MAINTENANCE COSTS	**30'575.12**	**3'459.00**	**12.16**	**9.42**	**8.48**	**N/A**	**N/A**	**N/A**	**N/A**
Local materials	18'208.95	2'060.00	7.24	5.61	5.05	N/A	N/A	N/A	N/A
Imported materials	-	-	0.00	-	-	N/A	N/A	N/A	N/A
Local equipments	-	-	0.00	-	-	N/A	N/A	N/A	N/A
Imported equipments	-	-	0.00	-	-	N/A	N/A	N/A	N/A
Labour	12'366.17	1'399.00	4.92	3.81	3.43	N/A	N/A	N/A	N/A
TOTAL OPERATION COSTS	**13'126.35**	**1'485.00**	**5.22**	**4.05**	**3.64**	**N/A**	**N/A**	**N/A**	**N/A**
Local materials	751.34	85.00	0.30	0.23	0.21	N/A	N/A	N/A	N/A
Imported materials	-	-	0.00	-	-	N/A	N/A	N/A	N/A
Local power services	-	-	0.00	-	-	N/A	N/A	N/A	N/A
Imported power services	-	-	0.00	-	-	N/A	N/A	N/A	N/A
Labour	12'375.01	1'400.00	4.92	3.81	3.43	N/A	N/A	N/A	N/A
TOTAL OTHER RECURRENT COSTS	**20'683.95**	**2'340.00**	**8.23**	**6.38**	**5.74**	**N/A**	**N/A**	**N/A**	**N/A**
Administration	20'683.95	2'340.00	8.23	6.38	5.74	N/A	N/A	N/A	N/A
Training	-	-	0.00	-	-	N/A	N/A	N/A	N/A
Promotion & education	-	-	0.00	-	-	N/A	N/A	N/A	N/A
TOTAL COST	**251'425.02**	**28'444.02**	**100**	**77.49**	**69.72**	**N/A**	**N/A**	**N/A**	**N/A**

	[inhabitant-year]/[life-cycle]	[household-year]/[life-cycle]	[litre/day-year]/[life-cycle]
PRESENT VALUE OF EXPECTED LIFE-CYCLE PRODUCTION	3'244	N/A	N/A

	[inhabitant-year]	[household-year]	[litre/day-year]
DESIGN CAPACITY	408	N/A	N/A

Costing summary

References

ADB-Asian Development Bank (1999). Handbook for the Economic Analysis of Water Supply Projects. Asian Development Bank, Manila, Philippines.

Bartram J. and Fewtrell L. (2001). Water Quality-Guidelines, Standards and Health: Assessment of Risk and Risk Management for Water-related Infectious Disease. IWA Publishing, London, UK.

Brikké F. and Bredero M. (2003). Linking Technology Choice with Operation and Maintenance in the Context of Community Water Supply and Sanitation: A Reference Document for Planners and Project Staff. WHO and IRC Water and Sanitation Centre, Geneva, Switzerland.

Brikké F., Bredero M., De Veer T. and Smet J. (1997). Linking Technology Choice with Operation and Maintenance for Low-cost Water Supply and Sanitation. WHO, Geneva, Switzerland; IRC International Water and Sanitation Centre, The Hague, Netherlands.

Cameron J., Hunter P., Jagals P. and Pond K. (2011). Valuing Water, Valuing Livelihoods. IWA Publishing, London, UK.

Carlevaro F. and Gonzalez C. (2011). Costing improved water supply systems for developing countries. *Water Management, Proceedings of the Institution of Civil Engineers*, **164**(WM3), 123–134.

Cumming O., Elliot M., Overbo A. and Bartram J. (2014). Does Global Progress on Sanitation Really Lag behind Water? An Analysis of Global Progress on Community- and Household-Level Access to Safe Water and Sanitation. *PLoS ONE*, **9**(12), e114699.

Deichmann U. (1999). Geographical Aspects of Inequality and Poverty. The World Bank, Washington DC, USA.

Dirección General de Programación Multianual del Sector Público (2006). Modifican el anexo SNIP-09-Parámetros de Evaluación. Resolución Directoral N° 006–2006-EF/68.01 (Modification of annex SNIP-09-Evaluation Parameters Directorial Resolution number 006–2006-EF/68.01). Official Gazette *El Peruano*, 9 September 2006.

FONCODES-Fondo de Cooperación para el Fondo Social (1995). Mapa de Inversión Social: FONCODES frente a la pobreza 1991–1994 (Mao of Social Investment: FONCODES against poverty 1991–1994), FONCODES, Lima, Peru.

Hutton G. and Haller L. (2004). Evaluation of the Costs and Benefits of Water and Sanitation Improvements at the Global Level, WHO-Protection of the Human Environment, Geneva, Switzerland.

INEI-Instituto Nacional de Estadística e Informática (2005). Censo Nacional de Vivienda 2005 (National Census of Housing 2005), INEI, Lima, Peru.

Johnson N. L., Kotz S. and Balakrishnan N. (1995). Continuous Univariate Distributions. John Wiley, New York, USA.

Lampoglia T. (2004). Evaluación de Costos para Sistemas de Agua y Saneamiento en el Ámbito Rural y de Pequeñas Localidades en el Perú (Cost Evaluation for Water and Sanitation Systems in Rural Areas and Small Localities in Peru), Report to the World Bank, Washington DC, USA, unpublished.

Lampoglia T. (2007). Testeo y Evaluación de la Metodología de Estimación de Costos de Proyectos No-convencionales de Agua y Saneamiento en el Perú (Testing and Evaluation of the Methodology for Estimating the Costs of Water and Sanitation Non-Conventional Projects in Peru), Report to the Department of Econometrics, University of Geneva, Geneva, Switzerland, unpublished.

Okun D. A. and Ernst W. R. (1987). Community Piped Water Supply Systems in Developing Countries: A Planning Manual, Technical Paper Number 60, The World Bank, Washington DC, USA.

Pannell D. J. and Schilizzi S. G. M. (2008). Economics and the Future: Time and Discounting in Private and Public Decision Making. Edward Elgar, Cheltenham, UK.

SKAT – Swiss Resource Centre and Consultancies for Development (2000–2001). Manuals on Drinking Water Supply, Volumes 1–7, SKAT, St. Gallen, Switzerland.

WHO (1993). Guidelines on Technologies for Water Supply Systems in Small Communities, CEHA Document N. TLM-5, WHO-Centre for Environmental Health Activities, Amman, Jordan.

WHO (2004). Water, Sanitation and Hygiene Links to Health. Facts and Figures. http://www.who.int/water_sanitation_ health/publications/factsfigures04/en/ (accessed 7 February 2015)

WHO (2008a). First Workshop on Costing of Improved Drinking-water Systems for Low-income Communities: Khon Kaen, 3–6 March 2008, report of an inter-regional workshop with participants from Cambodia, Indonesia, Lao PDR, Philippines, Thailand and Viet Nam, WHO-Protection of the Human Environment, Geneva, Switzerland. http://www.who.int/water_sanitation_health/economic/en/ (accessed 7 February 2015)

WHO (2008b). Second Workshop on Costing of Improved Drinking-water Systems for Low-income Communities: Luang Prabang, 29–31 October 2008, report of an inter-regional workshop with participants from Cambodia, Indonesia, Lao PDR, Philippines, Thailand and Viet Nam, WHO-Protection of the Human Environment, Geneva, Switzerland. http://www.who.int/water_sanitation_health/economic/en/ (accessed 7 February 2015)

WHO (2011). Guidelines for Drinking-water Quality, 4th edn. WHO Press, Geneva, Switzerland.

WHO/UNICEF (2000). Global Water Supply and Sanitation Assessment 2000 Report. WHO Press, Geneva, Switzerland.

WHO/UNICEF (2006). Drinking Water and Sanitation: A Compendium of Illustrations for Household Surveys. WHO, Geneva, Switzerland.

WHO/UNICEF (2010). Progress on Sanitation and Drinking-water 2010 Update. WHO Press, Geneva, Switzerland. http:// www.who.int/water_sanitation_health/publications/9789241563956/en/ (accessed 7 February 2015)

WHO/UNICEF (2012). Progress on Drinking Water and Sanitation: 2012 Update. WHO Press, Geneva, Switzerland and UNICEF Division of Communication, New York, USA. http://www.unicef.org/media/files/JMPreport2012.pdf (accessed 14 February 2015)

WHO/UNICEF (2013). Progress on Sanitation and Drinking-water – 2013 Update. WHO Press, Geneva, Switzerland. http://apps.who.int/iris/bitstream/10665/81245/1/9789241505390_eng.pdf (accessed 14 February 2015)

Young R. A. and Haveman R. H. (1985). Economics of Water Resources: A Survey. In: Handbook of Natural Resources and Energy Economics, Vol. II, A. V. Kneese and J. L. Sweeney (eds), Elsevier, Amsterdam-Lausanne-New York-Oxford-Shannon-Tokyo, pp. 465–529.

Zhuang J., Liang Z., Lin T. and De Guzman F. (2007). Theory and Practice in the Choice of Social Discount Rate for Cost-Benefit Analysis: A Survey, ERD Working Paper Series No. 94, Asian Development Bank, Manila, Philippines.

Glossary

All glossary entries are listed in strict alphabetical order and are set in boldface type.
Entry words used in the definition of others glossary entries are also set in boldface type for easy cross-referencing.

Activity of a water supply technology
The main activities performed by a water supply technology, to convey drinking-water from the water source to final users, are: **water collection**, **water conveyance**, **water storage**, **water treatment**, and **water distribution**. The proposed method of identifying, collecting and analysing cost data of **improved drinking-water supply technologies** relies on an analytical description of the water supply process based on these activities.

Activity-based costing method
This method, applied to a water supply project, is a cost accounting which breaks down the water supply process into a sequence of single activities each measured using quantitative indicators that are strongly correlated with the cost of the activity.

Aggregation level
Depending on the availability of the data sources, the costing of a water supply project can be performed at different levels of aggregation.

- An **item** aggregation level corresponds to the technologies that can be used to perform a single activity of an improved water supply system; for example, **water collection** can be carried out using different technologies, depending on the water source (such as a gutter for rainwater or a well for groundwater).
- A **sub-item** aggregation level corresponds to the particular technical device that can be used in practice to implement a technology; for example a wood gutter, a galvanized gutter or a PVC gutter can be used to collect rainwater.
- An **input** aggregation level corresponds to a breakdown of the sub-item description of a technical device, at a more detailed level at which cost data can be collected; for example a wood gutter structure and nails would be the (input level) **local materials** necessary to install a wood gutter for collecting rainwater.

Data collected at an item or sub-item aggregation level are recorded in the appropriate **water supply costing processor** spreadsheets by using the aggregated level option, while data collected at an input level are recorded using the disaggregated level option, both options being displayed on the north-east corner of the costs spreadsheets.

Average household size
This quantitative indicator, used to design (jointly with a **population scenario**) a **household scenario** for a water supply project, refers to the average number of inhabitants of the households benefiting from the project.

Average household size trend
The curve used to model the time evolution of the **average household size**.

Average incremental cost
Average incremental cost expresses a unit cost indicator of a water supply project (or of one of its components), namely the cost per unit of service provided by the project during its life-cycle. It is computed by dividing the **full cost present value** of the project (or of one of its components) by a measure of the project life-cycle production (see **life-cycle production present value**). Therefore, it may be viewed as the constant selling price of the services provided by a water supply project allowing to recover the total life-cycle costs of the project.

Average per capita consumption of water
This quantitative indicator, used to design (jointly with a **population scenario**) a **water scenario** for a water supply project, refers to the average number of litres of water per day supplied to the inhabitants benefiting from the project.

Average per capita consumption of water trend
The curve used to model the time evolution of the **average per capita consumption of water**.

Baseline scenario
A scenario showing, how the present situation would improve or deteriorate over the project **design lifetime** without implementation of the project.

Consumer price index
Index number used to transform historical cost data into monetary values at the **date of actualization** chosen for costing the project.

Costing questionnaires
These questionnaires have been designed to ease the identification and collection of data on costs. They describe the main **marketed resources** (materials, equipment, labour, power services, and so on) invested in the project at different **aggregation levels** (**item**, **sub-item** and **input** breakdown levels). They are based on a typology of costs reflecting the economic distinction between **investment costs** and **recurrent costs**.

Date of actualization
Reference date at which are valued and consolidated, all the cost components of the water supply project. This date is conventionally chosen to represent either the date of project completion or the beginning of the project life-cycle.

Design capacity/value
The maximum annual quantity of services, that can be provided by a given water supply project.

Design lifetime/period
The expected duration of the water supply project, expressed in years.

Discounting
The financial transaction by which, an amount of money available in the future, is priced in terms of today money using a **discount rate**.

Discount rate
The rate of interest used for computing the present value of an amount of money available in the future. The discount rate is:

- **annual**, when the **discounting** period is measured in years;
- **real**, when used for discounting the cost of resources by assuming no price-inflation of these resources;
- **social**, when used for discounting the costs of a social project.

Drinking-water
The WHO/UNICEF Joint Monitoring Programme has defined drinking-water as the water used for normal domestic purposes, including consumption and hygiene.

Free trend scenario
An option of the **water supply costing processor** that allows modelling the time evolution of a quantitative indicator used to design a **production scenario** of a water supply project by setting the value of the quantitative indicator for each year of the project life-cycle.

Full annual equivalent cost
The constant annuity to be paid during the project life-cycle to refund the **full cost present value** of a project at the same real annual **social discount rate** used for **discounting** the cost components of the full cost present value.

Full cost present value
A measure of the total life-cycle cost of a project obtained by consolidating all the cost components incurred during the project life-cycle, discounted at the **date of actualization** with a real **social discount rate**.

Household scenario
See **production scenarios**.

Household-year
A unit of measurement for the production of a water supply system expressing the services provided by the system to a household during a full year.

Imported materials
Materials imported from abroad.

Improved drinking-water source
The WHO/UNICEF Joint Monitoring Programme has classified **drinking-water** sources as either 'improved' or 'unimproved'. Improved drinking-water sources are those that by the nature of their construction adequately protect the source from outside contamination, in particular with faecal matter.

Improved drinking-water supply technologies
These technologies are those compatible with the Millennium Development Goals target of improving access to safe **drinking-water** in low-income communities. For operational purposes, they are identified

to the following **improved drinking-water sources**: piped water into dwelling, plot or yard; public tap or standpipe; tube well or borehole; protected dug well; protected spring; rainwater collection.

Improved water supply technology
See **improved drinking-water supply technologies**.

Incidental investment costs
These costs encompass labour costs for work supervision and engineering studies, as well as a lump sum for contingencies.

Incremental costing
The method of assessing the cost of a project intended to improve the current situation of water supply in a low-income community. It consists in computing the difference between the cost of supplying **drinking-water** with and without implementation of the project. The computation of an incremental cost requires complementing the **project scenario** with a **baseline scenario**.

Inhabitant-year
A unit of measurement for the production of a water supply system expressing the services provided by the system to an inhabitant during a full year.

Initial value
The value of a quantitative indicator used to design a **production scenario** of a water supply system, at the date that the project starts to be used.

Input
See **aggregation level**.

Investment costs
They encompass the costs of construction and equipment activities, including **preliminary studies** (technical, social and environmental aspects of the construction project), management, administration, coordination, logistics, transport, communications and office costs, as well as outsource tasks and quality control, along with any other unassigned and **incidental investment costs**.

Item
See **aggregation level**.

Least-cost analysis
A method aiming at identifying the most cost effective technological options to provide improved water supply to low-income communities. The analysis is conducted on the **locally appropriate technologies** by comparing the value of some cost indicator of the competing technologies, like the **average incremental cost**, the **full annual equivalent cost** or the **full cost present value.**

Life-cycle production present value
Measures the life-cycle production of a water supply system by valuing the services provided in the future less than those produced at the present time, just as costs incurred in the future have a lower present value than those incurred at the present time. Therefore, it expresses the present value of the total production over the life-cycle of a water supply facility if the unit value of the services provided remains constant.

Litre/day-year
A unit of measurement for the production of a water supply system expressing the litres of water per day supplied by the system during a full year.

Local currency
Currency used to value in monetary terms any cost component of the water supply project.

Locally appropriate technologies
Those technologies available for providing an improved **drinking-water** supply that are suited to the specific setting of a project. They are identified on the basis of a set of guiding criteria expressing the local constraints (water sources, financial, economic, technical, environmental, socioeconomic, socio-cultural, institutional, and so on) and health risks.

Local materials
Materials produced in the country where the project is implemented.

Maintenance costs
All the expenses for maintaining and repairing infrastructure, equipment and vehicles, in order to keep the water supply system in good condition during its life-cycle. In general, maintenance costs of equipment increase over time and they depend on the activity level. Therefore, to assess these costs it is important to have a **maintenance plan** for infrastructure and equipment, as formulated by the provider.

Maintenance plan
It identifies all the activities involved in maintaining and repairing the water supply facility, their levels for implementation (hours of work by activity, replacement parts, repairs procedures, inputs, and so on) and the timing or frequency at which these activities should be performed.

Marketed costs
The cost of the resources used in a water supply project that can be provided by markets.

Operation costs
Expenses for personnel, chemicals, electricity, fuel, materials, office supplies and building rents, required to keep a water supply system in operation during its life-cycle. In general, operation costs depend on the activity level. Therefore, to assess these costs it is important to design a **production scenario** of the life-cycle production growth of the water supply system.

Opportunity cost
The cost of any activity or resource measured in terms of the value of its best alternative use not chosen (forgone).

Opportunity cost of capital
The rate of payment on loans, which motivates an investor to borrow a marginal unit of capital to fund a productive activity that generates a higher future return (best alternative use).

Opportunity cost of spare capacity
It expresses the unit cost of the unused production capacity of a water supply system, namely the cost per unit of service provided by the system. It is valued as the difference between the **average incremental cost** of the system computed according to two variants of the use of its **design capacity**, namely the expected use and the full use of the design capacity throughout the system life-cycle. For a project designed to provide an increasing annual level of services, the average incremental cost computed by assuming a full use of the design capacity is expected to understate the average incremental cost for the expected use of the design capacity, leading to a positive value of the opportunity cost of spare capacity.

Other investment costs
Costs encompassing **preliminary studies**, administration, promotion, training and education/instruction of the project staff and users, as well as outsource tasks and quality control costs, along with any other unassigned **investment costs**.

Other recurrent costs
Cost encompassing activities necessary for the smooth functioning of the water supply system, in particular the costs for administration, training in administration, maintenance and operation, as well as health and hygiene promotion and education.

Population scenario
See **production scenarios**.

Population size
This quantitative indicator, used to design a **production scenario** of a water supply project, refers to the number of inhabitants benefiting from the project.

Population size trend
The curve used to model the time evolution of the **population size**.

Preliminary studies
Studies carried out at the pre-investment stage of a water supply project, to understand the technical, social and environmental aspects of the construction project.

Production scenarios
These scenarios define the time **trend shape** of the quantitative indicators used to design the life-cycle production growth of a water supply project. Such production scenarios can be designed, using the **water supply costing processor**, for:

- the size of the population supplied with water (**population scenario**), measured in terms of **inhabitant-year**;
- the number of household water connections (**household scenario**), measured in terms of **household-year**;
- the quantity of water supplied (**water scenario**), measured in terms of **litre/day-year**.

Project scenario
This scenario specifies the characteristics of the planned water supply project, like: location, timeframe, demand scenario, type of technology.

Project scenario questionnaire
This questionnaire, presented in Annex II, has been especially designed to collect basic information on the qualitative and quantitative characteristics of the project as a basis to formulate a project scenario.

Rate of time preference
It is the rate of return on loans, which motivates a consumer to save by postponing a marginal unit of current consumption in exchange for more future consumption.

Recurrent costs
These costs encompass **operation** and **maintenance costs**, as well as **other recurrent costs** like administrative costs, costs for training in administration, operation, maintenance, promotion and education to improve the hygiene habits in rural communities. Recurrent costs may be fix or variable with the level

of activity of the water supply facility. In the latter case, they a modelled as a function of one of the quantitative indicators used to design project life-cycle **production scenarios**.

Semi-skilled workers
These are workers between skilled and unskilled, depending on the type of work.

Shadow factor
It expresses the ratio of a **shadow** (economically efficient) **price** to its actual (economically inefficient) market price.

Shadow factor for foreign rate of exchange
Many governments do not permit free movement of their national currency in the international foreign exchange markets. They prefer to fix the value of their country's currency in relation to that of their major trading partner. As a result, the price for **imported materials** and equipment expressed in the national currency is below its international market value. The same result may also be achieved through a system of import restrictions. In such situations, the **shadow factor** for the foreign rate of exchange will be greater than one.

Shadow factor for unskilled labour wage
Because of minimum wage legislation, unskilled labour is economically overvalued, implying a wage higher than the real value of labour productivity, resulting in unemployment. Therefore, if there is widespread unemployment among the unskilled labour force in a country, the **shadow factor** for unskilled labour would be close to zero. In contrast, if a country has few unemployed **unskilled workers**, then this shadow factor would be close to one.

Shadow prices
When actual market prices involve significant distortions arising from market imperfections, they should be replaced by shadow prices assessing the cost of scarcity for the community of the marketed resources invested in a social project. These prices – including interest rates, wage rates and foreign exchange rates-should be those prevailing in a competitive economy.

Skilled workers
They include specialist civil engineers, civil constructors, technical staff, and social science professionals.

Social cost-benefit analysis
This analysis can be seen as an extension of the **social cost-effectiveness analysis**, aiming at comparing the social cost of the resources invested in a water supply project, to the complete set of its health and non-health social benefits.

Social cost-effectiveness analysis
This analysis seeks the best water supply project by comparing their social cost to their social health outcome. The project benefits on health are measured in terms of disability-adjusted life years (DALYs) averted, a broad measure of health taking in to account not only the additional years of life gained by the intervention but also the improved health that people enjoy as a consequence.

Social costing
The method of assessing the value of the **opportunity cost** of providing a given water supply service to the national economy. Therefore, this assessment must be conducted by looking at the costs of the project for the whole community and not just for a single agent (the private investor). In particular, **marketed costs** must be assessed using competitive market prices representing the national opportunity costs of

the resources invested in the project. Furthermore, it is advisable to account for non-market costs of the project, such as those generated by external effects or the use of public goods.

Social discount rate
See **discount rate**.

Social opportunity cost
The **opportunity cost** of a resource for the national economy.

Social opportunity cost of capital
Its use, as a discount rate for social projects, is advocated for those situations where public and private sectors compete for the same pool of funds and where social projects can inflict a loss to national consumption by diverting funds from more socially profitable private investments. In such situations, public investments should yield at least the same return as private investments. The social opportunity cost of capital can be approximated by the marginal pre-tax rate of return on riskless private investments. A relatively high social opportunity cost of capital is expected in developing countries, as a result of scarcity of capital, compared to that in developed countries.

Social rate of time preference
Its use, as a **discount rate** for social projects, is advocated on grounds that, unlike a private discount rate, the **social discount rate** should not merely express the average cost of capital invested in social projects but the inter-temporal substitution rate in consumption used to trade off the level of present national consumption against that of investments increasing future consumption.

Socioeconomic questionnaire
This questionnaire, presented in Annex II, has been specially designed to collect basic data on the demographic, economic and social characteristics of the community benefiting from the water supply project, as well as the potential contribution of the community to the project. The aim is to define the **baseline scenario**, and to help identify the socioeconomic constraints and risks faced by the project as inputs to the decision-making process, allowing for the identification and selection of the **locally appropriate technologies**.

Sub-item
See **aggregation level**.

Technical questionnaire
This questionnaire, presented in Annex II, has been specially designed to collect basic data on the urbanization, hydrological and climatic characteristics of the community benefiting from the water supply project. The aim is to define the **baseline scenario**, and to help identify the resources and technical constraints and risks faced by the project as inputs to the decision-making process, allowing for the identification and selection of the **locally appropriate technologies**.

Trend duration
The number of full years required, by a quantitative indicator used to design a **production scenario** of a water supply project, to reach its **design value** from its **initial value**.

Trend shape
The profile of the curve used to model the time evolution of a quantitative indicator used to design a **production scenario** of a water supply project. The **water supply costing processor** lists 6 pre-programmed profiles for the shape of the time trend.

Unskilled workers
These are manual workers that works, for example, in excavation, carrying of materials, or cleaning. In many instances, community members provide the workforce for these activities.

Valuation parameter
It is a datum requested by the 'Social costing' spreadsheet of the **water supply costing processor** to perform a **social costing** of the resources invested in the water supply project. It can be: the **date of actualization**, the **local currency**, the **shadow factors** for **unskilled labour wage** and **foreign rate of exchange**, or the real annual **social discount rate**.

Water collection
This main **activity of a water supply technology** requires the use of: an intake with pumping facilities, if the water is to be collected from a surface water source; a protected dug well, for the use of groundwater; a permanent roof, for the collection of rainwater.

Water conveyance
This main **activity of a water supply technology** is normally performed by gravity or pumping. In general, dedicated structures carry the water from the water source to a storage tank or reservoir before treatment or water consumption.

Water distribution
The means of performing this main **activity of a water supply technology** varies. **Drinking-water** may be piped or carried in containers by various means of transport by household members or middlemen (water vendors, tank carriers, and so on).

Water scenario
See **population scenarios**.

Water storage
This main **activity of a water supply technology** requires the use of reservoirs which have sufficient storage capacity for anticipated water demand before treatment and distribution.

Water supply costing processor
A specially developed Excel spreadsheet enabling a user-friendly identification, collection and processing of the relevant quantitative information to perform the **social costing** of the water supply projects.

Water treatment
The more common method to perform this main **activity of a water supply technology** includes sedimentation, aeration, filtration, demineralization and disinfection.

IWA Publishing's authorised EU representative for General Product Safety Regulations is Diane D'Arras, 15 rue Duret, 75116 Paris, France, e-mail: safety@iwap.co.uk.

Printed and bound by CPI Group (UK) Ltd, Croydon, CR0 4YY
12/05/2026
02108877-0004